Jim Baggott is an award-winning
scientist, he now works as an ind
maintains a broad interest in scie
continues to write on these subjects i

Farewell to Reality

Jim Baggott

WITHDRAWN

CONSTABLE • LONDON

Constable & Robinson Ltd
55–56 Russell Square
London WC1B 4HP
www.constablerobinson.com

First published in the UK by Constable,
an imprint of Constable & Robinson Ltd, 2013

A copy of the British Library Cataloguing in
Publication data is available from the British Library

ISBN: 978-1-78033-492-9 (paperback)
ISBN: 978-1-47210-470-0 (ebook)

Printed and bound in the UK

1 3 5 7 9 10 8 6 4 2

To John, in memory of Mary

Contents

CONTENTS

Preface

Modern physics is heady stuff. It seems that we can barely get through a week without being assaulted by the latest astounding physics story, its headlines splashed gaudily across the covers of popular science magazines and, occasionally, newspapers. The public's appetite for these stories is seemingly insatiable, and there's no escaping them. They are the subjects of innumerable radio and television news reports and documentaries, the latter often delivered with breathless exuberance and lots of arm-waving, from unconnected but always exotic locations, against a background of overly dramatic music.*

We might agree that these stories are all very interesting and entertaining. *But are they true?*

What evidence do we have for super-symmetric 'squarks', or superstrings vibrating in a multidimensional spacetime? How can we tell that we live in a multiverse? Is it really the case that the fundamental constituent at the heart of all matter and radiation is just 'information'? How can we tell that the universe is a hologram projected from information encoded on its boundary? What are we really supposed to make of the intricate network of apparent cosmic coincidences in the laws of physics?

Now, modern science has discovered that the reality of our physical existence is bizarre in many ways, but this is bizarreness for which there

---

* And, to my mind at least, produced for an audience of 12 year olds suffering from attention deficit disorder. Or maybe I'm just getting too old and cranky? Answers on a postcard please …

is an accumulated body of accepted scientific evidence. There is as yet *no* observational or experimental evidence for many of the concepts of contemporary theoretical physics, such as super-symmetric particles, superstrings, the multiverse, the universe as information, the holographic principle or the anthropic cosmological principle. For some of the wilder speculations of the theorists there can by definition *never* be any such evidence.

This stuff is not only not true, it is not even science. I call it 'fairy-tale physics'.* It is arguably borderline confidence-trickery.

Matters came to a head for me personally one evening in January 2011. That evening the BBC broadcast an edition of its flagship *Horizon* science series, entitled 'What is Reality?'. This began quite reasonably, with segments on the discovery of the top quark at Fermilab and some of the more puzzling conclusions of quantum theory. But beyond this opening the programme went downhill. It became a showcase for fairy-tale physics.

There was no acknowledgement that this was physics that had long ago lost its grip on anything we might regard as descriptive or explicative of the real world we experience. *Horizon* has an impressive reputation, and I became deeply worried that many viewers might be accepting what they were being told at face value. Conscious that I was now shouting rather pointlessly at my television, I decided that it was time to make a stand.

But, you might ask, what's the big deal? Why get so worked up? After all, consumers of popular science may simply wish to be entertained. They may wish to have their already boggled minds further boggled by the latest 'scientific' thinking, through a rapid succession of 'Oh wow!' revelations. Blimey! Parallel universes!

To take this view is, I believe, greatly to underestimate the people who consume popular science. It also shows an astonishing lack of respect. I suspect that many people might actually like to know what is accepted science fact and what is science fantasy. Only the hard facts can illuminate the situation sufficiently to make it possible to judge the nature of the trick, and to decide if it involves a betrayal of confidence, or even a betrayal of the truth.

---

* With acknowledgements to Piers Bizony.

PREFACE

Put it this way. If we were to regard fairy-tale physics as a lively branch of contemporary philosophy rather than science, do you think it would continue to receive the same level of attention from funding agencies, universities, popular science publishers, the producers of radio and television programmes and the wider public? No?

This is the big deal.

In writing this book, I've tried to hold on to several ambitions. I wanted to describe what modern physics has to say about the nature of our physical reality, based as far as possible on the accepted body of observationally or experimentally grounded scientific fact.

But we have to accept that even in this 'official' or 'authorized' version of reality there are many grey areas, where we run out of hard facts and have to deal with half-truths, guesses, maybes and a little imaginative speculation. This description is the nearest we can get to reality given the current gaps in our knowledge.

I also wanted to convince you that whilst the knowledge in this authorized version goes very deep, it does seem that we have paid a high price for it. We now know much more about the physical world than we have done at any other time in history. But I believe that we comprehend and understand much less.

We were obliged to abandon Isaac Newton's clockwork universe quite some time ago, but there was an inherent comprehensibility about this description that we found familiar and maybe even comforting (unless you happened to be a philosopher). The world according to quantum theory remains distinctly unfamiliar and uncomfortable. 'Nobody understands quantum mechanics,' declared the charismatic American physicist and Nobel laureate Richard Feynman, with some justification.* And today, more than a hundred years after it was first discovered, the theory remains completely inscrutable.

Some modern theoretical physicists have sought to compensate for this loss of understanding. Others have tried to paper over the cracks in theories that are clearly not up to the task. Or they have pushed, with

---

* This quote appears in Richard Feynman, *The Character of Physical Law*, (MIT Press, Cambridge, MA, 1965), p. 129.

vaulting ambition, for a final 'theory of everything'. These physicists have been led – unwittingly or otherwise – to myth creation and fairy tales.

I want to be fair to them. These physicists have been wrestling with problems for which there are as yet no observational or experimental clues to help guide them towards solutions. They are problem-rich, but data-poor. Rather than simply pleading ignorance or focusing their efforts on more tractable problems, they have chosen instead to abandon the obligation to refer their theories to our experience of the real world. They have chosen to abandon the scientific method.

In doing this, some theorists have railed against the constraints imposed by a scientific methodology that, they argue, has outlived its usefulness. They have declared that the time has come to embrace a new methodology for a 'post-empirical science'.

Or, if you prefer, they have given up.

With no observational or experimental data to ground their theories in reality, these theorists have been guided instead by their mathematics and their aesthetic sensibilities. Not surprisingly, ever more outrageous theoretical speculations freed from the need to relate to things happening in the world that we experience have transported us to the far wild shores of the utterly incredible and downright ridiculous.

This is not a wholly new phenomenon. Speculative theorising has always played an important role in scientific development, and in this book we will take a look at some examples from history. However, under the stark, unyielding gaze of the scientific method, in the light of new observational or experimental data such speculations have either become absorbed into mainstream science or they have fallen by the wayside and been rigorously forgotten.

But contemporary theoretical physics seems to have crossed an important threshold in at least two senses. Speculative theorizing of a kind that cannot be tested, that cannot be verified or falsified, a kind that is not subject to the mercilessness of the scientific method, is now almost common currency. The discipline has retreated into its own small, self-referential world. Its product is traded by its advocates as mainstream science within the scientific community, and peddled (or even missold) as such to the wider public.

Secondly, the unprecedented appetite for popular science and its attraction as an income stream have proved hard for the more articulate and eloquent of these advocates to resist. The result is that virtually every other popular book published on aspects of modern physics is chock-full of fairy stories. It is pseudo-science masquerading as science.

This will prove to be a controversial book. I'm not hopelessly naïve – I don't expect it to change current thinking or current practices. But I am hopeful that it will provoke some debate and, at the very least, provide a timely and much-needed antidote.*

In August 2011, I joined popular science writer Michael Brooks in a discussion about science at the Edinburgh International Book Festival. Brooks encouraged the assembled audience to imagine who members of the general public would name if asked to identify three scientists. He went on to suggest Albert Einstein, Stephen Hawking and Brian Cox.** The last is a former pop star turned high-energy physicist and television science presenter who has rapidly established himself as an important UK media personality. Interestingly, all are (or were) physicists.

The contrasting approaches of Einstein and Hawking are particularly relevant to the aims of this book. Hawking is a rather indulgent fairy-tale physicist, recently declaring that a form of untried and untested (and possibly untestable) superstring theory is the unified theory that Einstein spent the latter part of his life searching for.

Einstein, in contrast, was a theoretical physicist of the old school. His many pronouncements on the aims of science and the methods that scientists use are broadly consistent with common conception among both the majority of scientists and the wider public. On the basis of these pronouncements I suspect he would have been quite shocked by the state of contemporary theoretical physics. I had initially thought to title this book *What Would Einstein Say?* but have settled for trying to convey his likely sense of outrage by identifying for each chapter relevant quotations from his extraordinary lexicon.

---

* And yes, I'm also interested in the income-stream.
** My daughter, shortly to start her second year as a university drama student, was sitting dutifully in the audience. She thought about Brooks' question and challenged herself to name three scientists. Before Brooks could continue, she had identified Einstein, Hawking and Cox.

PREFACE

I am no longer a professional scientist, and some might argue that this means I am no longer qualified to hold an opinion on this subject. Obviously, I don't agree. I believe I have studied it long and hard enough to allow me not only to form and hold an opinion but also to express it as best I can.

But make no mistake, what you have in this book is an opinion that is very *personal*. Consequently, when I acknowledge a debt of gratitude to Professors Steve Blundell at Oxford University, Helge Kragh at Aarhus University in Denmark and Peter Woit at Columbia University in New York, who read and commented on the draft manuscript, I want to be absolutely clear that this acknowledgement should not suggest that they accept all my arguments. Of course, I take full responsibility for any errors or misconceptions that remain.

Jim Baggott
July 2012

1

The Supreme Task

Reality, Truth and the Scientific Method

The supreme task of the physicist is to arrive at those universal elementary laws from which the cosmos can be built up by pure deduction. There is no logical path to these laws; only intuition, resting on sympathetic understanding of experience, can reach them.

Albert Einstein[1]

Now I want to be absolutely clear and unequivocal upfront. I trained as a scientist, and although I no longer practise, I continue to believe – deeply and sincerely – that only science, correctly applied, can provide a sure path to true knowledge of the real world. If you want to know what the world is made of, where it came from, how it works and how it came to be as it is today, then my recommendation is to look to science for the answers.

I hope I speak with conviction, but be assured that I am not a zealot. I will happily admit that the practice of science is not always black and white. We are forced to admit shades of grey. It is a lot looser and more ambiguous than many practitioners are themselves often willing to admit. Much of the looseness and ambiguity arises because science is after all a human endeavour, and human beings are complicated and unpredictable things.

But it would be a mistake to think that the humanity of scientists is responsible for all the vagueness, that everything would be crystal clear if only a few flaky individuals would stick to the rules. When we look closely, we discover that what passes for the 'rules' of scientific endeavour are themselves rather vague and open to interpretation. This, I will argue, is how fairy-tale physics manages to thrive.

Our problems begin as soon as we try to unpack the sentences that I used to open this introductory chapter. Reality is at heart a metaphysical concept – it is, quite simply, 'beyond physics' and therefore beyond science. And just what, exactly, is this thing we call 'science'? For that matter, how should we define 'truth'?

That's a lot of difficult questions. And, it seems, if I'm going to accuse a bunch of theoretical physicists of abandoning the scientific method and so betraying the search for scientific truth about the nature of physical reality, then I'll need properly to ground this assertion in some definitions. It's better to try to clear all this up before we really get going.

There's quite a lot at stake here, so I've summarized my main conclusions about reality, science and truth in a series of six 'principles', handily picked out in italics with a grey background so that you can easily refer back to them if needed. Collectively, these principles define what it is that we apply science to, what science is and how we think we know when it is 'true'.

Of course, many physicists and philosophers of science will disagree with these principles, with varying degrees of vehemence. This, I think, is rather the point. What's important is how they seem to *you*.

In Part I, my mission will be to tell the story of the authorized version of reality in the context of these statements, showing how science has been applied to generate this contemporary version of the truth. This section concludes with a chapter summarizing most (but not all) of the problems with this authorized version and gives the reasons why we know it can't be the whole truth.

In Part II, I will attempt to explain how contemporary theoretical physics seeks to address these problems. It is here that fairy-tale physics sneaks in through unavoidable loopholes in our interpretation of one or more of the principles, but fails to satisfy all of them taken together. It is on this basis that I will seek to reject fairy-tale physics as metaphysics.

Let's start with reality.

Real is simply electrical signals interpreted by your brain

'What is real?' asked the character Morpheus in the 1999 Hollywood blockbuster movie *The Matrix*. 'How do you define real? If you're talking about what you can feel, what you can smell, what you can taste and see, then real is simply electrical signals interpreted by your brain.'[2]

These days we tend not to look for profundity in a Hollywood movie,* but it's worth pausing for a moment to reflect on this observation. I want to persuade you that reality is like liquid mercury: no matter how hard you try, you can never nail it down. I propose to explain why this is by reference to three 'everyday' things: a red rose, a bat and a dark cave.

So, imagine a red rose, lying on an expanse of pure white silk. We might regard the rose as a thing of beauty, its redness stark against the silk sheen of brilliant nothingness. What, then, creates this vision, this evocative image, this tantalizing reality? More specifically, what in reality creates this wonderful experience of the colour red?

That's easy. We google 'red rose pigment' and discover that roses are red because their petals contain a subtle mixture of chemicals called anthocyanins, their colour enhanced if grown in soil of modest acidity. So, anthocyanins in the rose petals interact with sunlight, absorbing certain wavelengths of the light and reflecting predominantly red light into our eyes. We look at the petals and we see red. This all seems quite straightforward.

But hang on. What, precisely, is 'red light'? Our instinct might be to give a scientific answer. Red light is electromagnetic radiation with wavelengths between about 620 and 750 billionths of a metre. It sits at the long-wavelength end of the visible spectrum, sandwiched between invisible infrared and orange.

But light is, well, light. It consists of tiny particles of energy which we call photons. And no matter how hard we look, we will not find an inherent property of 'redness' in photons with this range of wavelengths. Aside from differences in wavelength, there is nothing in the physical properties of photons to distinguish red from green or any other colour.

We can keep going. We can trace the chemical and physical changes that result from the interactions of photons with cone cells in your retina all the way to the stimulation of your visual cortex at the back of your brain. Look all you like, but you will not find the *experience* of the colour red in any of this chemistry and physics. It is obviously only

---

* Perhaps we've just lost the habit.

when you synthesize the information being processed by your visual cortex in your *conscious mind* that you experience the sensation of a beautiful red rose. And this is the point that Morpheus was making.

We could invent some equivalent scenarios for all of our other human senses – taste, smell, touch and hearing. But we would come to much the same conclusion. What you take to be your reality is just electrical signals interpreted by your brain.

What is it like to be a bat?

Now, you might be ready to dismiss all this as just so much juvenile philosophizing. Of course we're all reliant on the way our minds process the information delivered to us by our senses. But does it make any sense at all for the human mind to have evolved processes that represent reality differently from how it really is? Surely what we experience and the way we experience it *must* correspond to whatever it is that's 'out there' in reality? Otherwise how could we survive?

To answer these questions, it helps to imagine what it might be like to be a bat.

What does the world look like – what passes for reality – from a bat's point of view? We know that bats compensate for their poor night vision by using sophisticated sonar, or echolocation. Bats emit high-frequency sounds, most of them way above the threshold of human perception. These sound waves bounce off objects around them, forming echoes which they then detect.

Human beings do not use echolocation to gather information about the world. We cannot possibly imagine what it's like for a bat to be a bat because we lack the bat's sensory apparatus, in much the same way that we cannot begin to describe colours to someone who has been blind from birth.

But the bat is a highly evolved mammal, successful in its own ecological niche. Just because I can't understand what reality might be like for a bat doesn't mean that the bat's perceptions and experiences of that reality are any less legitimate than mine.

What this suggests is that evolutionary selection pressures lead to the development of a sensory apparatus that delivers a finely tuned *representation* of reality. All that matters is that this is a representation that lends a creature survival advantages. There is no evolutionary selection pressure to develop a mind to represent reality as it really is.

Plato's allegory of the cave

So, what do we perceive if not reality as it really is? In *The Republic*, the ancient Greek philosopher Plato used an allegory to describe the situation we find ourselves in. This is his famous allegory of the cave.

Imagine you are a prisoner in a dark cave. You have been a prisoner all your life, shackled to a wall. You have never experienced the world outside the cave. You have never seen sunlight. In fact, you have no knowledge of a world outside your immediate environment and are not even aware that you are a prisoner, or that you are being held in a cave.

It is dark in the cave, but you can nevertheless see men and women passing along the wall in front of you, carrying all sorts of vessels, and statues and figures of animals. Some are talking. As far as you are concerned, the cave and the men and women you can see constitute your reality. This is all you have ever known.

Unknown to you, however, there is a fire constantly burning at the back of the cave, filling it with a dim light. The men and women you can see against the wall are in fact merely shadows cast by real people passing in front of the fire. The world you perceive is a world of crude appearances of objects which you have mistaken for the objects themselves.

Plato's allegory was intended to show that whilst our reality is derived from 'things-in-themselves' – the real people that walk in front of the fire – we can only ever perceive 'things-as-they-appear' – the shadows they cast on the cave wall. We can never perceive reality for what it is; we can only ever perceive the shadows. '*Esse est percipi*', declared the eighteenth-century Irish philosopher George Berkeley: essence is perception, or to be is to be perceived.

These kinds of arguments appear to link our ability to gain knowledge of our external reality firmly with the workings of the human mind. A disconnect arises because of the apparent unbridgeable distance between the physical world of things and the ways in which our perception of this shapes our mental world of thoughts, images and ideas. This disconnect may arise because we lack a rigorous understanding of how the mind works. But knowing how the mind works wouldn't change the simple fact that thoughts are very different from things.

Veiled reality

It doesn't end here. Another disconnect, of a very different kind but no less profound, is that between the quantum world of atomic and subatomic dimensions and the classical world of everyday experience. What we will discover is that our anxiety over the relationship between reality and perception is extended to that between reality and measurement.

Irrespective of what thoughts we think and how we think them, we find that we can no longer assume that what we *measure* necessarily reflects reality as it really is. We discover that there is also a difference between 'things-in-themselves' and 'things-as-they-are-measured'.

The contemporary physicist and philosopher Bernard d'Espagnat called it 'veiled reality', and commented that:

> … we must conclude that physical realism is an 'ideal' from which we remain distant. Indeed, a comparison with conditions that ruled in the past suggests that we are a great deal more distant from it than our predecessors thought they were a century ago. [3]

At this point the pragmatists among us shrug their shoulders and declare: 'So what?' I can never be sure that the world as I perceive or measure it is really how the world is 'in reality', but this doesn't stop me from making observations, doing experiments and forming theories about it. I can still establish facts about the shadows – the projections of reality into our world of perception and measurement – and I can compare these with similar facts derived by others. If these facts agree, then surely we have learned something about the nature of the reality that lies beneath the shadows. We can still determine that if we do *this*, then *that* will happen.

Just because I can't perceive or measure reality as it really is doesn't mean that reality has ceased to exist. As American science-fiction writer Philip K. Dick once observed: 'Reality is that which, when you stop believing in it, doesn't go away.'[4]

And this is indeed the bargain we make. Although we don't always openly acknowledge it upfront, 'reality-in-itself' is a metaphysical concept. The reality that we attempt to study is inherently an *empirical reality* deduced from our studies of the shadows. It is the reality of observation, measurement and perception, of things-as-they-appear and of things-as-they-are-measured. As German physicist Werner Heisenberg once claimed: '… we have to remember that what we observe is not nature in itself but nature exposed to our method of questioning'.[5]

But this isn't enough, is it? We may have undermined our own confidence that there is anything we can ever know about reality-in-itself, but we must still have some *rules*. Whatever reality-in-itself is really like, we know that it must exist. What's more, it must surely exist independently of perception or measurement. We expect that the shadows would continue to be cast whether or not there were any prisoners in the cave to observe them.

We might also agree that, whatever reality is, it does seem to be rational and predictable, within recognized limits. Reality appears to be logically consistent. The shadows that we perceive and measure are not completely independent of the things-in-themselves that cause them. Even though we can never have knowledge of the things-in-themselves, we can *assume* that the properties and behaviour of the shadows they cast are somehow determined by the things that cast them.

That feels better. It's good to establish a few rules. But don't look too closely. If you want some assurance that there are good, solid scientific reasons for believing in the existence of an independent reality, a reality that is logical and structured, for which our cause-and-effect assumptions are valid, then you're likely to be disappointed. To repeat one last time, reality is a metaphysical concept – it lies beyond the grasp of science. When we adopt specific beliefs about reality, what we are actually doing is adopting a specific *philosophical position*.

If we accept the rules as outlined above, then we're declaring ourselves as *scientific realists*. We're in good company. Einstein was a realist, and when asked to justify this position he replied: 'I have no better expression than the term "religious" for this trust in the rational character of reality and in its being accessible, to some extent, to human reason.'[6]

Now, it's one thing to be confident about the existence of an independent reality, but it's quite another to be confident about the existence of overtly theoretical entities that we might want to believe to exist in some shape or form within this reality. When we invoke entities that we can't directly perceive, such as photons or electrons, we learn to appreciate that we can't know anything of these entities as things-in-themselves. We may nevertheless choose to assume that they exist. I can find no better argument for such 'entity realism' than a famous quote from philosopher Ian Hacking's book *Representing and Intervening*. In an early passage in this book, Hacking explains the details of a series of experiments designed to discover if it is possible to reveal the fractional electric charges characteristic of 'free' quarks.* The experiments involved studying the flow of electric charge across the surface of balls of superconducting niobium:

> Now how does one alter the charge on the niobium ball? 'Well, at that stage,' said my friend, 'we spray it with positrons to increase the charge or with electrons to decrease the charge.' From that day forth I've been a scientific realist. *So far as I'm concerned, if you can spray them then they are real.*[7]

This brings us to our first principle.

> **The Reality Principle**. *Reality is a metaphysical concept, and as such it is beyond the reach of science. Reality consists of things-in-themselves of which we can never hope to gain knowledge. Instead, we have to content ourselves with knowledge of empirical reality, of things-as-they-appear or things-as-they-are-measured. Nevertheless, scientific realists assume that reality (and its entities) exists objectively and independently of perception or measurement. They believe that reality is rational, predictable and accessible to human reason.*

Having established what we can and can't know about reality, it's time to turn our attention properly to science.

---

* Alas, it's not possible.

The scientific method

In 2009, Britain's Science Council announced that after a year of deliberations, it had come up with a definition of science, perhaps the first such definition ever published: 'Science is the pursuit of knowledge and understanding of the natural and social world following a systematic methodology based on evidence.'[8]

Given that any simple definition of science is likely to leave much more unsaid than it actually says, I don't think this is a bad attempt. It all seems perfectly reasonable. There's just the small matter of the 'systematic methodology', the cold, hard, inhuman, unemotional logic engine that is supposed to lie at the very heart of science. A logic that we might associate with Star Trek's Spock.

The 'scientific method' has at least three components. The first concerns the processes or methodologies that scientists use to establish the hard facts about empirical reality. The second concerns methods that scientists use to create abstract theories to accommodate and explain these facts and make testable predictions. The third concerns the methods by which those theories are tested and accepted as true or rejected as false. Let's take a look at each of these in turn.

Getting at the facts

The first component seems reasonably straightforward and should not detain us unduly. Scientists pride themselves on their detachment and rigour. They are constantly on the lookout for false positives, systematic errors, sample contamination, anything that might mislead them into reporting empirical facts about the world that are later shown to be wrong.

But scientists are human. They are often selective with their data, choosing to ignore inconvenient facts that don't fit, through the application of a range of approaches that, depending on the circumstances, we might forgive as good judgement or condemn as downright fraud. They make mistakes. Sometimes, driven by greed or venal ambition, they might cheat or lie.

There is no equivalent of a Hippocratic oath for scientists, no verbal or written covenant to commit them to a system of ethics and work solely for the benefit of humankind. Nevertheless, ethical behaviour is deeply woven into the fabric of the scientist's culture. And the emphasis on repetition, verification and critical analysis of scientific data means that any mistakes or wrongdoing will be quickly found out.

Here's a relevant example from contemporary high-energy physics. The search for the Higgs boson at CERN's Large Hadron Collider has involved the detection and analysis of the debris from trillions upon trillions of protons colliding with each other at energies of seven and, most recently, eight trillion electron volts.* If the Higgs boson exists, then one of the many ways in which it can decay involves the production of two high-energy photons, a process written as H $\rightarrow \gamma\gamma$, where H represents the Higgs boson and the Greek symbol γ (gamma) represents a photon.

About three thousand physicists have been involved in each of two detector collaborations searching for the Higgs, called ATLAS and CMS.** One of their tasks is to sift through the data and identify instances where the proton–proton collisions have resulted in the production of two high-energy photons. They narrow down the search by looking for photons emitted in specific directions with specific energies. Even so, finding the photons can't be taken as evidence that they come from a Higgs boson, as theory predicts that there are many other ways in which such photons can be produced.

The physicists therefore have to use theory to calculate the 'background' events that contribute to the signal coming from the two photons. If this can be done reliably, and if any systematic errors in the detectors themselves can be estimated or eliminated, then any significant excess events can be taken as evidence for the Higgs.

On 21 April 2011, an internal discussion note from within the ATLAS collaboration was leaked to a high-energy physics blogger. The note suggested that clear evidence for a Higgs boson had been found in the H $\rightarrow \gamma\gamma$ decay channel, with a signal thirty times greater than predicted.

---

* That's a lot of trillions, which is why the Large Hadron Collider cost £5 billion to construct. By the way, an electron volt is the amount of energy a single negatively charged electron gains when accelerated through a one-volt electric field. A 100W light bulb burns energy at the rate of about 600 billion billion electron volts per second.

** ATLAS stands for A Toroidal LHC Apparatus. CMS stands for Compact Muon Solenoid.

If this was true, it was fantastic, if puzzling, news. But it wasn't true. The purpose of internal discussion notes such as this is to allow the exchange of data and analysis within the collaboration before a collective, considered view is made public. It was unfortunate that the note had been leaked. Within just a few weeks, ATLAS released an official update based on the analysis of twice as much collision data as the original note, work that no doubt demanded many more sleepless nights for those involved. There was no excess of events. No Higgs boson – yet.[*]

As ATLAS physicist Jon Butterworth subsequently explained:

> Retaining a detached scientific approach is sometimes difficult. And if we can't always keep clear heads ourselves, it's not surprising people outside get excited too. This is why we have internal scrutiny, separate teams working on the same analysis, external peer review, repeat experiments, and so on.[9]

This was a rare example in which the public got to see the way science self-regulates, how it uses checks and balances in an attempt to ensure that it gets its facts right. Scientists don't really like us looking over their shoulders in this way, as they fear that if we really knew what went on, this would somehow undermine their credibility and authority.

I take a different view. The knowledge that science can be profoundly messy on occasion simply makes it more human and accessible; more Kirk than Spock. Knowing what can go wrong helps us to appreciate that when it does go seriously wrong, this is usually an exception, rather than the rule.

No facts without theory

The process of building a body of accepted scientific facts is often fraught with difficulty, and rarely runs smoothly. We might be tempted to think that once we have built it, this body of evidence forms a clear, neutral, unambiguous substrate on which scientific theories can be

---

[*] We'll look more closely at the search for the Higgs boson in Chapter 3.

contrived. Surely the facts form a 'blank sheet of paper', on which the theorists can exercise their creativity?

But this is not the case. It is in fact impossible to make an observation or perform an experiment without the context of a supporting theory in some shape or form. French physicist and philosopher Pierre Duhem once suggested that we go into a laboratory and ask a scientist performing some basic experiments on electrical conductivity to explain what he is doing:

> Is he going to answer: 'I am studying the oscillations of the piece of iron carrying this mirror?' No, he will tell you that he is measuring the electrical resistance of a coil. If you are astonished, and ask him what meaning these words have, and what relation they have to the phenomena he has perceived and which you at the same time perceived, he will reply that your question would require some long explanations, and he will recommend that you take a course in electricity.[10]

Facts are never theory-neutral; they are never free of contamination from some theory or other. As we construct layer upon layer of theoretical understanding of phenomena, the concepts of our theories become absorbed into the language we use to describe the phenomena themselves. Facts and theory become hopelessly entangled.

If you doubt this, just look back over the previous paragraphs concerning the search for the Higgs boson at CERN.

This brings us to our second principle.

The Fact Principle. *Our knowledge and understanding of empirical reality are founded on verified scientific facts derived from careful observation and experiment. But the facts themselves are not theory-neutral. Observation and experiment are simply not possible without reference to a supporting theory of some kind.*

So how do scientists turn this hard-won body of evidence into a scientific theory?

Theory from facts: anything goes?

The naïve answer is to say that theories are derived through a process of *induction*. Scientists use the data to evolve a system of generalizations, built on abstract concepts. The generalizations may be elevated to the status of natural patterns or 'laws'. The laws in turn are explained as the logical and inevitable result of the properties and behaviour of a system of theoretical concepts and theoretical entities.

A suitable example appears to be provided by the German mathematician and astronomer Johannes Kepler, who deduced his three laws of planetary motion after years spent mulling over astronomical data collected by the eccentric Dane Tycho Brahe, at Benatky Castle and Observatory near Prague.

Brahe's painstaking observations of the motion of the planet Mars suggested a circular orbit around the sun, to within an accuracy of about eight minutes of arc. But this was not good enough for Kepler:

> ... if I had believed that we could ignore these eight minutes, I would have patched up my hypothesis accordingly. But since it was not permissible to ignore them, those eight minutes point the road to a complete reformulation of astronomy.[11]

Brahe's observations were just too good.

In his book *Astronomia Nova* (*New Astronomy*), published in 1609, Kepler used Brahe's data to argue that the planets move not in circular orbits around the sun, but in elliptical orbits with the sun at one focus. For this scheme to work, he had to assume that the earth behaves like just any other planet, also moving around the sun in an orbit described by an ellipse.*

This means that a planet moves closer to the sun for some parts of its orbit and further away for other parts. Kepler also noted a balance between the distance of the planet from the sun and the speed of its motion in the orbit. A planet moves faster around that part of its orbit that takes it closest to the sun, and more slowly in that part of its orbit

---

* Although Copernicus had argued for a sun-centred planetary system, the debate was still raging in Kepler's time. Brahe himself argued for a system in which the planets orbit the sun, which, in turn, orbits a stationary earth.

that is more distant. An imaginary line drawn from the sun to the planet will sweep out an area as the planet moves in its orbit. Kepler deduced that the balance between speed and proximity to the sun means that no matter where the planet is in its orbit, such an imaginary line will sweep out equal areas in equal times.

In 1618, Kepler added a third law. The cube of the mean radius of the orbit divided by the square of the period (the time taken for a planet to complete one trip around the sun) is approximately constant for all the planets in the solar system.

Kepler had used Brahe's facts to develop a set of empirical laws based on the abstract concept of an elliptical orbit. In 1687, Isaac Newton deepened our understanding by devising a theory that explained the origins of Kepler's elliptical orbits in terms of other abstract concepts – the forces acting between bodies – in three laws of motion and a law of universal gravitation.

So, this seems all very straightforward. Kepler induced his three laws from Brahe's extensive set of accurate astronomical data. Newton then 'stood on the shoulders of giants', using Kepler's conclusions, among many others, to derive his own laws, thus driving science inexorably in the direction of ultimate truth. Right?

Wrong. Whenever historians examine the details of scientific discoveries, they inevitably find only confusion and muddle, vagueness and error, good fortune often pointing the way to the right answers for the wrong reasons. Occasionally they find true genius. Theorizing involves a deeply human act of creativity. And this, like humour, doesn't fare well under any kind of rational analysis.

Kepler himself was a mystic. He was pursuing a theory of the solar system in which the sun-centred orbits of the six planets then known were determined by a sequence of Platonic solids, set one inside the other like a Russian matryoshka doll. In arriving at his three laws, he assumed that the planets were held in their orbits by magnetic-like emanations from the sun, which dragged them around like spokes in a wheel. He inverted the roles of gravity and inertia. He made numerous basic arithmetic errors which happened to cancel out.

He stumbled on the mathematical formula that describes the planetary orbits but did not realize that this represented the equation for an ellipse. In frustration, he rejected this formula and, growing increasingly convinced that the orbit must be an ellipse instead, tried that, only to discover that it was the same formula that he'd just abandoned. 'Ah, what a foolish bird I have been,' he wrote.[12]

Not to worry. Newton figured it all out 69 years later, clearing the confusion, sorting the muddle, clarifying the vagueness and eradicating the errors. Except that he didn't. Although there can be no doubting that Newton's mechanics represented a tremendous advance, in truth he had simply replaced one mystery with another.

Kepler's elliptical orbits were no longer a puzzle. They could be explained in terms of the balance between gravitational attraction and the speed with which a planet moves in its orbit. We were left instead to puzzle over the origin of Newton's force of gravity, which appeared to act instantaneously, at a distance, and with no intervening medium to carry it. Newton himself lamented:

> That gravity should be innate, inherent, and essential to matter, so that one body may act upon another, at a distance through a vacuum, without the mediation of anything else, by and through which their action and force may be conveyed from one to another, is to me so great an absurdity, that I believe no man who has in philosophical matters a competent faculty of thinking, can ever fall into it.[13]

He was accused of introducing 'occult elements' into his mechanics. This mystery would not be resolved for another two hundred years, when Einstein replaced Newton's gravity with the interplay between matter and curved spacetime.

What this excursion into the history of science tells us is that while there can be no doubt at all that Kepler and Newton developed theories and made discoveries that represented real advances in our knowledge of empirical reality, the methods by which these advances were made do not lend themselves to a simple, convenient rationalization. It seems odd that we can't devise a 'scientific' theory about how science is actually supposed to work, a universal scientific method applicable to all science for all time. But it's a fact.

As Einstein himself admitted: 'There is no logical path to these laws; only intuition, resting on sympathetic understanding of experience, can reach them.'

Now intuition is all about the acquisition of knowledge without inference or the use of logical reasoning. Often jammed in the door of intuition we find the foot of speculation. If it is the case that in the act of scientific creativity 'anything goes', then the door is wedged firmly open for all manner of speculative theorising.*

The importance of the abstract

When a theory is constructed, it will contain concepts that are more or less familiar, often depending on the degree of speculation involved. One important point to note is that the concepts that form the principal base ingredients of any scientific theory are utterly abstract.

For example, in the standard model of particle physics we find elementary particles such as electrons and quarks. These are already rather abstract conceptual entities, but they are not the ones I'm referring to here. Underpinning the theory on which the standard model is based is the abstract mathematical concept of the *point particle*. This is an idealization, an assumption that for mathematical convenience particles like electrons and quarks can be represented as though they have no spatial extension, with all their mass concentrated to an infinitesimally small point.

Now when the theory is used to make predictions on distance scales much larger than the particles themselves, the assumption of point particles is not likely to cause much of a problem. But as we probe ever smaller distance scales, we can expect to run into trouble. String theory was developed, in part, as a way of avoiding problems with point particles. In this case, the abstract mathematical concept of a zero-dimensional point particle is replaced by another abstract mathematical concept of a one-dimensional string (and, subsequently, many-dimensional 'branes').

These mathematical abstractions form a kind of 'toolkit' that theorists use to construct their theories. It is a toolkit of points, limits, asymptotes,

---

* For a vivid account of some of the more extreme methods that scientists have used to make discoveries, see Michael Brooks, *Free Radicals: The Secret Anarchy of Science*, (Profile Books, London, 2011).

infinities, infinitesimals, and much more.* There is little or nothing we can do about this abstraction, but we must not forget it is there. It will have some very important implications for scientific methodology.

> **The Theory Principle**. *Although physical theories are constructed to describe empirical facts about reality, they are nevertheless founded on abstract mathematical (we could even say metaphysical) concepts. The process of abstraction from facts to theories is highly complex, intuitive and not subject to simple, universal rules applicable to all science for all time. In the act of scientific creation, any approach is in principle valid provided it yields a theory that works.*

The Theory Principle begs some obvious questions. Specifically, given the complex and intuitive nature of scientific creation, how are we supposed to know if a theory 'works'?

Putting theories to the test

This, at least, appears to have a straightforward answer. We know a theory works because we can *test* it. Now this may be a test against pre-existing facts – observations or experimental data – that are necessarily different from the facts from which the theory was sprung. Or it may involve a whole new set of observations or a new series of experiments specifically designed for the purpose.

There are plenty of examples from history. Einstein's general theory of relativity correctly predicted observations of the advance in the perihelion (the point of closest approach to the sun) of the planet Mercury – a pre-existing fact that couldn't be explained using Newton's theory of universal gravitation. The general theory of relativity also predicted that the path of light from distant stars should be bent in the vicinity of sun, a prediction borne out by observations during a solar

---

* For a recent, highly readable tour through the abstract, see Giovanni Vignale, *Beautiful Invisible: Creativity, Imagination, and Theoretical Physics*, (Oxford University Press, 2011).

eclipse in May 1919, recorded by an expedition led by British astrophysicist Arthur Eddington.*

It is when we try to push beyond the test itself in search of a more specific criterion for scientific methodology that we start to run into more problems. In the 1920s, a group of philosophers that came to be known as the Vienna Circle sought to give verification pride of place. For a theory to be scientific, they argued, it needs to be *verifiable* by reference to the hard facts of empirical reality. This sounds reasonable, until we realize that this leaves us with no certainty. As British philosopher Bertrand Russell put it:

> But the real question is: Do any number of cases of a law being fulfilled in the past afford evidence that it will be fulfilled in the future? If not, it becomes plain that we have no ground whatever for expecting the sun to rise tomorrow … It is to be observed that all such expectations are only *probable*; thus we have not to seek for a proof that they must be fulfilled, but only for some reason in favour of the view that they are likely to be fulfilled.[14]

This is a life-or-death issue, as Russell went on to explain: 'The man who has fed the chicken every day throughout its life at last wrings its neck instead, showing that more refined views as to the uniformity of nature would have been useful to the chicken.'[15]

Verifiability won't do, argued Austrian philosopher Karl Popper. He suggested *falsifiability* instead. Theories can never be verified in a way that provides us with certainty, as the chicken can attest, but they can be falsified. A theory should be regarded as scientific if it can in principle be falsified by an appeal to the facts.

But this won't do either. To see why, let's take another look at an episode from the history of planetary astronomy.

The planet Uranus was discovered by William Herschel in 1781.** When Newton's mechanics were used to predict what its orbit should

---

* Eddington was also selective with his data, with history rewarding his choices as 'good judgement'.
** How you pronounce 'Uranus' is entirely up to you. I'm old enough to be stuck with the pronunciation I learned at school: 'your anus'. This does not make me titter, or blush, because I am no longer eight years old.

be, this was found to disagree with the observed orbit. What happened? Was this example of disagreement between theory and observation taken to falsify the basis of the calculations, and hence the entire structure of Newtonian mechanics?

No, it wasn't.

Remember that theories are built out of abstract mathematical concepts, such as point particles or gravitating bodies treated as though all their mass is concentrated at their centres. If we think about how Newton's laws are actually applied to practical situations, such as the calculation of planetary orbits, then we are forced to admit that no application is possible without a whole series of so-called *auxiliary* assumptions or hypotheses. And, when faced with potentially falsifying data, the tendency of most scientists is not to throw out an entire theoretical structure (especially one that has stood the test of time), but instead to tinker with the auxiliary assumptions.

This is what happened in this case. The auxiliary assumption that was challenged was the (unstated) one that the solar system consists of just seven planets. British astronomer John Adams and French mathematician Urbain Le Verrier independently proposed that this assumption be abandoned in favour of the introduction of an as yet unobserved eighth planet that was perturbing the orbit of Uranus. In 1846 the German astronomer Johann Galle discovered the new planet, subsequently called Neptune, less than one degree from its predicted position.

This does not necessarily mean that a theory can never be falsified, but it does mean that falsifiability is not a robust criterion for a scientific method. Emboldened by his success, in 1859 Le Verrier challenged the same auxiliary assumption in attempting to solve the problem of the anomaly in the perihelion of Mercury. He proposed another as yet unobserved planet – which he called Vulcan – between the sun and Mercury itself.

No such planet could be found. When confronted by potentially falsifying data, either the theory itself or at least one of the auxiliary assumptions required to apply it must be modified, but the observation or experiment does not tell us which. In fact, in this case it was Newton's theory of universal gravitation that was at fault.

So, neither verifiability nor falsifiability provides a sufficiently robust criterion for defining 'science'. And yet history shows that some theories have indeed been falsified and that others have been at least temporarily 'verified', in the sense that they have passed all the tests that have been thrown at them so far.

It seems to me that the most important defining criterion is therefore the *testability* of the theory. Whether we seek to verify it or falsify it, and irrespective of what we actually do with the theory once we know the test results, to qualify as a scientific theory it should in principle be testable.

The Testability Principle. *The principal requirement of a scientific theory is that it should in some way be testable through reference to existing or new facts about empirical reality. The test exposes the veracity or falsity of the theory, but there is a caveat. The working stuff of theories is itself abstract and metaphysical. Getting this stuff to apply to the facts or a test situation typically requires wrapping the abstract concepts in a blanket of auxiliary assumptions; some explicitly stated, many taken as read. This means that a test is rarely decisive. When a test shows that a theory is false, the theory is not necessarily abandoned. It may simply mean that one or more of the auxiliary assumptions are wrong.*

I want to be clear that the demand for testability in the sense that I'm using this term should not be interpreted as a demand for an immediate yes-no, right-wrong evaluation. Theories take time to develop properly, and may even be perceived to fail if subjected to tests before their concepts, limitations and rules of application are fully understood. Think of testability instead as more of a professional judgement than a simple one-time evaluation.

The Testability Principle demands that scientific theories be actually or potentially capable of providing tests against empirical facts. Isn't this rather loose? How can we tell if a novel theoretical structure has the *potential* for yielding predictions that can be tested? For sure, it would be a lot easier if this was all black and white. But I honestly don't think it's all that complicated. A theory which, despite considerable effort, shows absolutely no promise of progressing towards testability should not be regarded as a scientific theory. A theory that continually fails repeated tests is a failed theory.

The correspondence theory of truth

All this talk of verification and falsification implies that 'truth' plays a central role in whatever we conclude passes for a scientific methodology. But according to the Reality Principle, the reality we seek to study is a metaphysical concept. The Theory Principle says that the common currency of scientific theories involves yet more metaphysics.

It seems obvious that concepts that start out as purely speculative cannot be considered to be true. By the same token, concepts accepted as true may later turn out to be false. Phlogiston was once believed to be a substance contained in all combustible materials, released when these materials burned. This is not what we believe today.

How should we reconcile all this with the idea of scientific truth?

What convinces us either way is, of course, *evidence* in the form of facts. Here, we anchor the idea of 'truth' firmly to the idea of an empirical reality. A statement is true if, and only if, it corresponds to established facts about the real world derived from our perceptions or measurements.

This *correspondence theory* of truth implies that there can be no truth — no right or wrong — without reference to facts about the external world.* True statements tell it like it really is. The correspondence theory is the natural choice of a realist.

We appreciate that scientific theories are imperfect in many ways and, as Russell noted, we cannot assume that what we regard as true today will necessarily be true tomorrow. But we can think of scientific theories as possessing a *verisimilitude*, a truth-likeness, which increases with each successive generation of scientific development.

Scientists are constantly refining and improving their theories, and as it makes no sense to develop new theories that are less useful than their predecessors, we might be prepared to accept that successive developments help take us closer and closer to 'the truth' about empirical reality.

The correspondence theory shifts the burden from the meaning of 'truth' to the meaning of 'facts', which seems both logical and helpful.

---

* There is an alternative *coherence* theory of truth which asserts that truth is determined by relations between statements rather than correspondence to external facts. We will return to this alternative interpretation of truth in Chapter 10.

Most often, scientists who claim to be in pursuit of the truth are actually in pursuit of the facts. An interesting and unexpected experimental result will prompt a flurry of activity in competing laboratories, as scientists rush to confirm or refute the facts and so establish the truth.

For example, on 4 July 2012, scientists at CERN declared that they had discovered a new particle 'consistent' with the standard-model Higgs boson. Further research will be required to categorize fully its properties and behaviour. This research will establish empirical facts about the existence of a particle that was first hypothesized in 1964. If these facts emerge broadly as anticipated, we may conclude that the statement 'the Higgs boson exists in nature' is true, in much the same way that 'unicorns exist in nature' or 'phlogiston exists in nature' are false.

The Veracity Principle. *It is not possible to verify a scientific theory such that it provides absolute certainty for all time. A theory is instead accepted (or even tolerated), on the basis of its ability to survive the tests and meet additional criteria of simplicity, efficacy, utility, explanatory power and less rational, innately human measures such as beauty. Over time the theory becomes familiar and is accepted as 'true' or, at least, as possessing a high truth-likeness or verisimilitude. The scientists' confidence or degree of belief in the theory grows. It is eventually absorbed into the common body of knowledge which forms the current 'authorized' version of empirical reality.*

The Copernican attitude

There is one final, critically important ingredient to consider. When scientists go about their business – observing, experimenting, theorizing, predicting, testing and so on – they tend to do so with a certain fixed attitude or mindset. It is this attitude that sets science apart, that lends it its 'Spock-ness' and exposes it to occasional criticism as a soulless, somewhat inhuman enterprise.

Let me explain.

The purpose of an organized system of religion is to enable its followers to come to terms with their place in the universe, give meaning to their lives and offer moral instruction and comfort in times of need. Religion is all about 'us'. It puts us at the centre of things, and although not all religious systems necessarily claim that the external physical world is organized principally for our benefit, many do.

Science is very different. Scientists tend to assume that there is, in fact, nothing particularly special about 'us'. We are not uniquely privileged observers of the universe we inhabit. We are not at the centre of everything. There is nothing special about the planet on which we exist. Or the rather average class G2 main-sequence star that gives us sunlight. Or the galaxy of between 200 and 400 billion stars in which our sun orbits, about two thirds or 25,000 light years from the centre. Or the 53 other galaxies which together with the Milky Way form the Local Group. Or the Virgo Supercluster of which the Local Group forms part. I could go on, but I think you've got the message.

This is the Copernican Principle.

The Copernican Principle. *The universe is not organized for our benefit and we are not uniquely privileged observers. Science strives to remove 'us' from the centre of the picture, making our existence a natural consequence of reality rather than the reason for it. Empirical reality is therefore something that we have learned to observe with detachment, without passion. Scientists ask fundamental questions about how reality works and seek answers in the evidence from observation and experiment, irrespective of their own personal preferences, prejudices and beliefs.*

The principle is misnamed insofar as Nicolaus Copernicus himself did not view his heliocentric model of the universe as necessarily undermining earth's unique position. What he offered was a technical improvement over the Ptolemaic system's obsession with convoluted structures constructed using epicycles. By putting the sun at the centre, the retrograde motions of the other planets in the solar system could be explained as apparent motions when observed from an orbiting (rather than stationary) earth. The planets don't really move backwards: they only appear to move backwards because we are also being transported around the sun.

But Copernicus sparked a revolution that was to shape the very nature of science itself. It became apparent that science works best when we remove 'us' as the primary objective or purpose of the equations of reality.

Nearly five hundred years on, all the scientific evidence gathered thus far suggests that the Copernican Principle is justified. The evidence suggests rather strongly that we are not privileged: we are not at the centre of things. It points to the singular absence of an intelligent force, moving in mysterious ways to organize the world just for our benefit. As French mathematician and astronomer Pierre-Simon, marquis de Laplace, once advised Napoleon: '*Je n'avais pas besoin de cette hypothèse-là.*'*

Now this has been a bit of a whirlwind tour through some aspects of the philosophy of science and scientific methodology. I hope it wasn't too much like hard work.

I've tried to be reasonable. This 'working model' of science acknowledges that reality-in-itself is metaphysical, that the objects of scientific study are the shadows, the things-as-they-appear or things-as-they-are-measured. It accepts that the facts that scientists work with are not theory-neutral – they do not come completely free from contamination by theoretical concepts. It accepts that theories are in their turn populated by metaphysical concepts and mathematical abstractions and are derived by any method that works, from induction to the most extreme speculation. It acknowledges that theories can never be accepted as the ultimate truth. Instead, they are accepted as possessing a high truth-likeness or verisimilitude – they correspond to the facts. In this way they become part of the authorized version of empirical reality.

Finally, the model acknowledges the important role played by the Copernican attitude. Science works best when we resist the temptation to see ourselves as the primary objective or purpose of reality.

This is a bit more elaborate than the Science Council's definition. But this elaboration is more reflective of actual practice. It is necessary if we are to understand how fairy-tale physics is not only possible, but is able to thrive.

---

* 'I had no need of that hypothesis.' From W. W. Rouse Ball, *A Short Account of the History of Mathematics*, (4th edition, 1908).

Part I

The Authorized Version

White Ambassadors of Morning

Light, Quantum Theory and the Nature of Reality

The more success the quantum theory has, the sillier it looks. How non-physicists would scoff if they were able to follow the odd course of developments!

Albert Einstein[1]

Two years before they hit the stratosphere with *Dark Side of the Moon* in 1973, the British progressive rock band Pink Floyd released their sixth studio album, called *Meddle*. Side two is a single 23-minute-long track called 'Echoes'.* After a short, discordant nightmare sequence in the middle of the track, 'Echoes' greets the morning with a return to harmony. Soft, gentle vocals declare: 'And through the window in the wall, comes streaming in on sunlight wings, a million white ambassadors of morning.'[2]

This is a particularly pleasant metaphor for photons, the elementary particles that constitute all electromagnetic radiation, including light. However, on a bright morning there are rather more than a million of them streaming through the window. And photons are no ordinary particles. For one thing, they have no mass. They have *spin* – an intrinsic angular momentum – which we perceive as polarisation. They also have something called *phase*, which means that they are particles that can also behave like waves.

Let's look at them a bit more closely.

---

* Or track 6 if you're not familiar with the structure of a long-playing record.

Einstein's light quantum hypothesis

In fact, photons were the first 'quantum particles', suggested by Einstein in a paper he published in 1905. At that time, light was thought to consist of a series of wave disturbances, with peaks and troughs moving much like the ripples that spread out on the surface of a pond where a stone has been thrown.

The evidence for wave behaviour was very compelling. Light can be diffracted. When forced to squeeze through a narrow aperture or slit in a metal plate, it spreads out (diffracts) in much the same way that ocean waves will spread if forced through a narrow gap in a harbour wall. This is behaviour that's hard to explain if light is presumed to be composed of particles obeying Newton's laws of motion and moving in straight lines.

Light also exhibits interference. Shine light on two narrow apertures or slits side by side and it will diffract through both. The wave spreading out beyond each aperture acts as though it comes from a 'secondary' source of light, and the two sets of waves run into each other. Where the peak of one wave meets the peak of the other, the result is constructive interference – the waves mutually reinforce to produce a bigger peak. Where trough meets trough the result is a deeper trough. But where peak meets trough the result is destructive interference: the waves cancel each other out.

The result is a pattern of alternating brightness and darkness called *interference fringes*, which can be observed using photographic film. The bright bands are produced by constructive interference and the dark bands by destructive interference. This is called two-slit interference.

The wave model of light was given a compelling theoretical foundation in a series of papers published in the 1860s by Scottish physicist James Clerk Maxwell. He devised an elaborate model which combined the forces of electricity and magnetism into a single theory of electromagnetism. Maxwell's theory consists of a complex set of interconnecting differential equations, but can be greatly simplified for the case of electromagnetic radiation in a vacuum. When recast, these equations look exactly like the equations of wave motion. Maxwell himself discovered that the speed of these 'waves of electromagnetism' is predicted to be precisely the speed of light.

But waves are disturbances *in* something, and it was not at all clear what light waves were meant to be disturbances in. Some physicists (including Maxwell) argued that these were waves in a tenuous form of matter called the ether, which was supposed to pervade the entire universe. But subsequent experimental searches for evidence of the ether came up empty.

Einstein didn't believe that the ether existed, and argued that earlier work in 1900 by German physicist Max Planck hinted at an altogether different interpretation. He boldly suggested that Planck's result should be taken as evidence that light consists instead of independent, self-contained 'bundles' of energy, called *quanta*.

> According to the assumption considered here, in the propagation of a light ray emitted from a point source, the energy is not distributed continuously over ever-increasing volumes of space, but consists of a finite number of energy quanta localized at points of space that move without dividing, and can be absorbed or generated only as complete units.[3]

This was Einstein's 'light quantum hypothesis'. He went on in the same paper to predict the outcomes of experiments on something called the photoelectric effect. This effect results from shining light on to the surfaces of certain metals. Light with wave frequencies above a threshold characteristic of the metal will cause negatively charged electrons to be kicked out from the surface. This was a bit of a challenge for the wave theory of light, as the energy in a classical wave depends on its intensity (related to the wave amplitude, the height of its peaks and depth of its troughs), not its frequency. The bigger the wave, the higher its energy.*

Einstein figured that if light actually consists of self-contained bundles of energy, with the energy of each bundle proportional to the frequency of the light, then the puzzle is solved.[4] Light quanta with low frequencies don't have enough energy to dislodge the electrons. As the frequency is increased, a threshold is reached above which the absorption of a light quantum knocks an electron out of the lattice of metal ions at the surface. Increasing the intensity of the light simply

---

* Think about the destructive forces unleashed when the energy contained in a tsunami strikes land.

increases the number (but not the energies) of the light quanta incident on the surface. He went on to make some simple predictions that could be tested in future experiments.

These were highly speculative ideas, and physicists did not rush to embrace them. Fortunately for Einstein, his work on the special theory of relativity, published in the same year, was better regarded. It was greeted as a work of genius.

When in 1913 Einstein was recommended for membership of the prestigious Prussian Academy of Sciences, its leading members – Planck among them – acknowledged his remarkable contributions to physics. They were prepared to forgive his lapse of judgement over light quanta:

> That he may sometimes have missed the target in his speculations, as, for example, in his hypothesis of light-quanta, cannot be really held against him, for it is not possible to introduce really new ideas even in the most exact sciences without sometimes taking a risk.[5]

The risk was partly rewarded just two years later. When American physicist Robert Millikan reported the results of further experiments on the photoelectric effect, Einstein's predictions were all borne out. The results were declared to be supportive of the predictive ability of Einstein's equation connecting photoelectricity and light frequency, but the light quantum hypothesis remained controversial.

Einstein was awarded the 1921 Nobel Prize for physics for his work on the photoelectric effect (but not the light quantum). Two years later, American physicist Arthur Compton and Dutch theorist Pieter Debye showed that light could be 'bounced' off electrons, with a predictable change in light frequency. These experiments appear to demonstrate that light does indeed consist of particles moving in trajectories, like small projectiles. Gradually, the light quantum became less controversial and more acceptable. In 1926, the American chemist Gilbert Lewis coined the name 'photon' for it.

Wave particle duality and the Copenhagen interpretation

But this couldn't simply be a case of reverting to a purely particulate description of light. Nobody was denying all the evidence of wave-like behaviour, such as diffraction and interference. Besides, Einstein had retained the central property of 'frequency' in his description of the light quanta, and frequency is a property of waves.

So, how could particles of light also be waves? Particles are by definition localized bits of stuff – they are 'here' or 'there'. Waves are delocalized disturbances in a medium; they are 'everywhere', spreading out beyond the point where the disturbance is caused. How could photons be here, there *and* everywhere?

Einstein believed that photons are first and foremost particles, following predetermined trajectories through space. In this scheme, the trajectories are determined by some other, unguessed property that leads to the *appearance* of wave behaviour as a result of statistical averaging. According to this interpretation, in the two–slit interference experiment each photon follows a precise and predetermined trajectory. It is only when we have observed a large number of photons that we see that the trajectories bunch together in some places and avoid other places, and we interpret this bunching as interference.

There was another view, however. Danish physicist Niels Bohr and German Werner Heisenberg argued that particles and waves are merely the shadowy projections of an unfathomable reality into our empirical world of measurement and perception. They claimed that it made no sense to speculate about what photons *really are*. Better to focus on how they *appear* – in this kind of experiment they appear as waves, in that kind of experiment they appear as particles.

Bohr is credited with the statement:

> There is no quantum world. There is only an abstract quantum physical description. It is wrong to think that the task of physics is to find out how nature is. Physics concerns what we can say about nature.[6]

This approach to quantum theory became known as the *Copenhagen interpretation*, named for the city in which Bohr had established a physics institute where much of the debate about the interpretation of quantum theory took place. At the heart of this interpretation lies Bohr's notion of *complementarity*, a fundamental duality of wave and particle behaviour.

Photon spin and polarization

Before we continue with this exploration of the historical development of quantum theory, and what it tells us about the ways in which we try to understand light, it's important to take a short diversion to look more closely at the properties of photons.

Photons are massless particles which in a vacuum travel at the speed of light. The Particle Data Group, an international collaboration of some 170 physicists that reviews data on elementary particles and produces an annual 'bible' of recommended data for practitioners, suggests that photons cannot possess a so-called 'rest mass' greater than about two millionths of a billionth of a billionth (2×10^{-24}) of the rest mass of the electron.* If the photon does have a rest mass, it is indeed very, *very* small.

Photons are also electrically neutral – they carry no electrical charge. They are instead characterized by their energies, which are directly related to their frequencies or inversely to their wavelengths. This reference to wave-like behaviour, with properties of frequency and wavelength, belies a property called *phase* that all photons (and in fact all types of quantum particle) possess.

In one sense, the phase of a wave is simply related to the position it has in its peak and trough cycle. However, in practical terms, the phase of a quantum particle can never be observed directly. Measurements reveal properties that are affected by the phase, but not the phase itself. According to the Copenhagen interpretation, we should deal only with what we can measure. Whatever the origin of phase, it results in the observation of behaviour that we interpret in terms of waves, and we must leave it at that.

Photons are also characterized by their intrinsic angular momentum, or what physicists call *spin*. In classical Newtonian mechanics, we associate angular momentum with objects that spin around some central axis. For a solid body spinning on its axis, such as a spinning top or the earth, the angular momentum is calculated from its moment of

---

* See http://pdg.lbl.gov. Click 'Summary Tables' and select the top entry 'Gauge and Higgs Bosons (gamma, g, W, Z, …)'. The photon is here referred to as 'gamma'. The rest mass is the mass that a photon would have if it could be (hypothetically) slowed down and stopped.

inertia (a measure of the body's resistance to rotational motion) multiplied by the speed of the rotation.

This obviously can't be applicable for photons. For one thing, photons are massless: they have no central axis they can spin around. They can't be made to spin faster or slower. The term 'spin' is actually quite misleading. It is a hangover from an early stage in the development of quantum theory, when it was thought that this property could be traced to an intrinsic 'self-rotation', literally quantum particles spinning like tops. This was quickly dismissed as impossible, but the term 'spin' was retained.

As with phase, it doesn't help to look too closely at the property of spin and ask what a photon is really doing. We do know that the property of spin is manifested as angular momentum. The interactions between photons and matter are governed by the conservation of angular momentum, and experiments performed over many years have demonstrated this. When we create an intense beam of photons (such as a laser beam), selected so that the spins of the photons are aligned, the angular momentum of all the individual photons adds up and the beam imparts a measurable torque. A target placed in the path of the beam will be visibly twisted in the direction of the beam's rotation.

Physicists characterize the spin properties of quantum particles according to a *spin quantum number*, which provides a measure of the particles' intrinsic angular momentum.[7] This quantum number can take half-integral or integral values. Quantum particles with half-integral spins are called *fermions*, named for Italian physicist Enrico Fermi. If we persist in pushing the spinning top analogy, we find that fermions would have to spin twice around their axes in order to get back to where they started (which shows that persisting with classical analogies in the quantum world usually leads only to headaches).

The most important thing to know about fermions is that they are forbidden from occupying the same quantum 'state'. In this context the state of a quantum particle is defined in terms of the various properties the particle possesses, such as energy and angular momentum. These properties are characterized by their quantum numbers, so a quantum state is defined in terms of its particular set of quantum numbers. Thus when we say that fermions are forbidden from occupying the same quantum state, we mean that no two fermions can have the same set of quantum numbers.

This is called the Pauli exclusion principle, first devised by Austrian physicist Wolfgang Pauli in 1925, and it explains the existence of all material substance and the structure of the periodic table of the elements.

The photon has a spin quantum number of 1, which classifies it as a *boson*, named for Indian physicist Satyendra Nath Bose. Unlike fermions, photons can 'condense' into a single quantum state in which many photons possess the same quantum numbers. A laser is just one striking example of such 'Bose condensation'. It is not possible to create a 'laser' using electrons instead of photons, since electrons are not bosons and so cannot be formed in a single quantum state in this way.

Although the spin of a quantum particle is a rather mysterious property, we do know that it can 'point' in different directions. Generally speaking, the spin of a quantum particle with spin quantum number s can point in a number of directions given by twice the value of s plus 1. For electrons, s is equal to ½, so there are two times ½ plus 1, or two different spin directions. We tend to call these 'spin up' and 'spin down'.

This would suggest that the spin of the photon (spin quantum number 1) can point in three different directions (two times 1 plus 1), but there's a caveat. The spins of particles that travel at the speed of light cannot point in the direction they're travelling.

The reasons for this are both subtle and complex. But let's suppose for a minute that we observe a spinning top moving in a straight line across a table. Suppose further that from our vantage point above the spinning top we observe it to be spinning clockwise as it moves to the right (in the direction of three o'clock). Now, the total speed of a single point on the leading edge of the top is the vector sum of the rotation speed at that point plus the speed of linear motion across the table. So, as the top rotates from three o'clock to nine o'clock, the total speed is a little less than the speed of linear motion. But as the top rotates from nine o'clock back to three o'clock, the total speed is a little greater than the speed of linear motion. And there's the rub. If the object is a photon moving at the speed of light, then, as we will discover in Chapter 4, Einstein's special theory of relativity insists that this is a speed that cannot be exceeded. A photon 'rotating' such as to give a total speed slightly greater than the speed of light is physically unacceptable and is indeed forbidden by the theory. This reduces the number of possible spin directions for the photon to two.

These two spin 'orientations' of the photon correspond to the two known types of circular polarization. In right–circular polarization the amplitude of the photon wave can be thought to rotate clockwise as seen from the perspective of the source of the photon. In left-circular polarization the photon wave rotates counterclockwise. When viewed in terms of rotating waves, it's perhaps not hard to appreciate the connection between spin and angular momentum.

If circular polarization is unfamiliar to you, don't worry. Waves are very malleable things. They can be combined in ways that particles cannot. If we don't like one kind of wave, we can add other kinds to it to form something called a *superposition*. Add left- and right-circular polarized waves together in just the right ways and you get linear polarization – vertical and horizontal – which is much more familiar.* Photons in different polarization states have been used since the 1970s in some of the most profound tests of the interpretation of quantum theory ever performed.

Quantum probability and the collapse of the wavefunction

Newtonian physics is characterized by a determinism founded on a strong connection between cause and effect. In the old version of empirical reality described by Newton's theories, if I do *this*, then *that* will happen. No question. One hundred per cent.

In quantum theory this kind of predictability is lost. Think about what happens in a two-slit interference experiment when the intensity of the light is reduced so low that, on average, only one photon passes through the apparatus at a time. What we see is that each photon is detected on the other side of the slits, perhaps as a tiny white dot formed on a piece of photographic film. If we wait patiently for lots and lots of photons to pass through the apparatus, one at a time, then we will observe that the white dots form an interference pattern (see Figure 1).

How does this work? The photon is single quantum of light. It is an indivisible particle, unable to split in two and pass through both slits simultaneously. But waves can do just this, and the photon is also a wave. A wave is described mathematically by something called a *wavefunction*. To explain what happens in this case, we assume that the wavefunction corresponding to a single photon passes through both

---

* Physicists call this changing the *basis* of the description.

slits. The secondary wavefunctions emerging from the two slits then diffract and interfere. By the time they reach the photographic film, the wavefunctions have combined to produce bright fringes (constructive interference) and dark fringes (destructive interference).

We now make a further important assertion. The amplitude of the wavefunction at a particular point in space and time provides a measure of the *probability* that a photon is present.[*] When the wavefunction corresponding to a single photon interacts with chemicals in the photographic emulsion, it 'collapses'. At this point the photon mysteriously appears, as an indivisible bundle of energy, and a white dot is formed on the film. Such dots are more likely to be formed in regions of the photographic plate where the amplitude of the wavefunction is high, less likely where it is low. After many photons have been recorded, the end result is a set of interference fringes.

We need to be very careful here. Quantum probability is not like 'ordinary' probability, of the kind we associate with tossing a coin. When I toss a coin, I know that there's a 50 per cent probability that it will land 'heads' and a 50 per cent probability that it will land 'tails'. I don't know what result I'm going to get for any specific toss because I'm ignorant of all the variables involved – the weight of the coin, speed of the toss, air currents, the force of the coin's impact on the ground, and so on. If I could somehow acquire knowledge of some of these variables and eliminate others completely, then I might actually be able to use Newton's laws of motion to compute in advance what result I'm going to get.

Quantum probability is quite different. The Copenhagen interpretation insists that in a quantum system like the two-slit interference experiment with single photons, there are no other variables of which we are ignorant. There is nothing in quantum theory that tells us how an indivisible photon particle navigates its way past the two slits. This doesn't necessarily mean that we're missing something; that the theory is somehow incomplete. What it does mean is that the particle picture is not relevant here and we can't use it to understand

---

[*] Actually, the probability is related to the modulus-square of the amplitude. Amplitudes can be positive or negative or 'imaginary' (i.e. they depend on i, the square root of -1), but, by definition, probabilities are always positive.

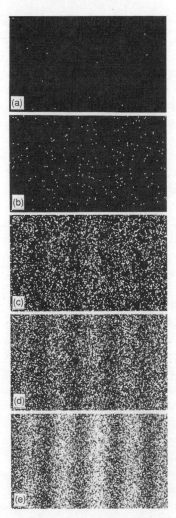

Figure 1 We can observe quantum particles as they pass, one at a time, through a two-slit apparatus by recording where they strike a piece of photographic film. Each white dot indicates that 'a quantum particle struck here'. Photographs (a)–(e) show the resulting images when, respectively, 10, 100, 3,000, 20,000 and 70,000 particles have been detected. The interference pattern becomes more and more visible as the number of particles increases. From A. Tonomura et al., *American Journal of Physics, 57 (1989)*, pp. 117–20.

what's going on. We use the wave picture instead and revert to the particle picture only when the wavefunction collapses and the photon is detected.

'We have to remember that what we observe is not nature in itself but nature exposed to our method of questioning,' Heisenberg said.

This notion of the collapse of the wavefunction doesn't completely break the connection between cause and effect, but it does weaken it considerably. In the quantum domain if I do *this*, then *that* will happen with a certain probability. No certainty. Some doubt. Quantum events like the detection of a photon appear to be left entirely to chance.

Einstein didn't like it at all:

> Quantum mechanics is very impressive. But an inner voice tells me that it is not yet the real thing. The theory produces a good deal but hardly brings us closer to the secret of the Old One. I am at all events convinced that *He* does not play dice.[8]

The collapse of the wavefunction continues to plague the current authorized version of reality.

The uncertainty principle

According to Newton's mechanics, although I might not have the means at my disposal, I can have some confidence that I can measure with arbitrary precision the position and momentum of an object moving through space.

Imagine we fire a cannonball. The cannonball shoots out of the cannon with a certain speed and traces a parabolic path through the air before hitting the ground. It moves through space, passing instantaneously through a series of positions with specific speeds. Although here again I would need to figure out all the different variables at play (wind speed, the precise pull of earth's gravity), there is nothing in principle preventing me from finding out what these are. After some computation, I produce a map of position and speed throughout the cannonball's trajectory.

If I know the mass of the cannonball (which I can measure separately), I can calculate the momentum at each point in the trajectory, by multiplying together mass and speed at that point.

Once again, however, in quantum mechanics things are rather different. Quantum particles are also waves. Suppose we were somehow able to localize a quantum wave particle in a specific region of space so that we could measure its position with arbitrary precision. In the wave description, this is in principle possible by combining a large number of wave forms of different frequencies in a superposition (called a 'wavepacket'), such that they add up to produce a resultant wave which is large in one location in space and small everywhere else. Great. This gives us the position.

What about the momentum? That's a bit of a problem. We localized the wave by combining lots of waves with different frequencies. This means that we have a spread of frequencies in the superposition and hence a spread of wavelengths.[9] According to French physicist Louis de Broglie, the wavelength of a quantum wave is inversely proportional to the quantum particle's momentum.[10] The spread of wavelengths therefore means there's a spread of momenta.

We can measure the position of a quantum wave particle with arbitrary precision, but only at the cost of uncertainty in the particle's momentum.

The converse is also true. If we have a quantum wave particle described by a single wave with a single frequency, this implies a single wavelength which we can measure with arbitrary precision. From de Broglie's relation we determine the momentum. But then we can't localize the particle. It remains spread out in space. We can measure the momentum of a quantum wave particle with arbitrary precision, but only at the cost of uncertainty in the particle's position.

This is Heisenberg's famous uncertainty principle, which he discovered in 1927.[11]

Heisenberg initially interpreted his principle in terms of what he thought of as the unavoidable 'clumsiness' with which we try to probe the quantum domain with our essentially classical measuring instruments. Bohr had come to a different conclusion, however, and they argued bitterly. The clumsiness argument implied that quantum wave particles actually possess precise properties of position and momentum, and we could in principle measure these if only we had the wit to devise experiments of greater subtlety.

Bohr was adamant that these properties simply do not exist in our empirical reality. This is a reality that consists of things-as-they-are-measured – the wave shadows or the particle shadows, as appropriate. Bohr insisted that it is this fundamental duality, this complementarity of wave and particle behaviour, that lies at the root of the uncertainty principle, much as the explanation given above suggests. It is not possible for us to conceive experiments of greater subtlety, because such experiments are inconceivable.

Heisenberg eventually bowed to the pressure. He accepted Bohr's view and the Copenhagen interpretation was born.

The uncertainty principle is not limited to position and momentum. It applies to other pairs of physical properties, called *conjugate* properties, such as energy and time. It also applies to the different spin orientations of quantum particles.

For example, photon polarization can be 'vertical' or 'horizontal', which implies some kind of reference frame against which we judge these orientations. 'Vertical' must mean vertical with regard to some co-ordinate axis. When applied to polarization, the uncertainty principle tells us that certainty in one co-ordinate axis means complete uncertainty in another. If in the laboratory I fix a piece of Polaroid film so that its transmission axis lies along the z axis (say), and I measure photons passing through this film, then I have determined that these photons have vertical polarization measured along the z axis, with a high degree of certainty.* This implies a high degree of uncertainty for polarizations oriented along either the x or y axis.

Quantum fluctuations of the vacuum

The science fiction writer Arthur C. Clarke famously formulated three laws of prediction. The third law, a guide for aficionados of 'hard science fiction' (characterized by its emphasis on scientific accuracy), declares that any sufficiently advanced technology is indistinguishable from magic.**

---

* This is only a high degree of certainty rather than absolute certainty as the polarizing film won't be perfect. It will allow some photons that aren't precisely vertically polarized to 'leak' through.

** I'm sure you want to know what the other two laws are. The first law says that when a distinguished but elderly scientist states that something is possible, he is almost certainly right. When he states that something is impossible, he is almost certainly wrong. The second law says that the only way to discover the limits of the possible is to venture a little way beyond them into the impossible.

So, here's an interesting bit of quantum physics that at first glance looks indistinguishable from magic.

Take two small metal plates and place them side by side a few millionths of a metre apart in a vacuum, insulated from any external electric and magnetic fields. There is no force between these plates, aside from an utterly insignificant gravitational attraction between them which, for the purposes of this experiment, can be safely ignored.

Now here comes the magic. Although there can be no force between them, the plates are actually pushed very slightly together.

This is an effect first identified by the Dutch physicist Hendrik Casimir in 1948. Experiments conducted in 2001 by a team of physicists at Italy's National Institute of Nuclear Physics and the Department of Physics at the University of Padua demonstrated the existence of a 'Casimir force' in this specific arrangement of two closely spaced metal plates.

What's going on?

Heisenberg's uncertainty principle sets a fundamental threshold which nothing in nature can cross. We might be used to thinking about the uncertainty principle in relation to constraints placed on the position and momentum of a quantum particle, or on its energy and the rate of change of this energy with time. We might not think to apply the uncertainty principle to the vacuum; to 'empty' space. What would be the point?

Suppose we create a perfect vacuum, completely insulated from the external world. We might be tempted to argue that there is 'nothing' at all in this vacuum. But what does this imply? It implies that the energy of an electromagnetic field in the vacuum is zero. It also implies that the rate of change of the amplitude of this field is zero too. But the uncertainty principle denies that we can know precisely the energy of an electromagnetic field and its rate of change. They *can't* both be exactly zero.

What happens is that the vacuum suffers a bad case of the jitters. It experiences fluctuations of the electromagnetic field which average out to zero, in terms of both energy and rate of change, but which are nevertheless non-zero at individual points in space and time.

Fluctuations in a quantum field are equivalent to quantum particles. The vacuum fluctuations of the electromagnetic field can be thought of as *virtual photons*; 'virtual' not because they are not 'real', but because they are not directly perceived.

41

In the experiment devised by Casimir, the space between the plates constrains the types of vacuum fluctuations that can persist. Only fluctuations that 'fit' in this space can contribute to the vacuum energy. Alternatively, we can imagine that the narrow space between the plates reduces the number of virtual photons that can persist there. The density of virtual photons between the plates is then lower than the density of virtual photons elsewhere. The end result is that the plates experience a kind of virtual 'radiation pressure'; the higher density of virtual photons on the outsides of the plates pushes them closer together.

The Italian physicists' painstaking measurements showed that the magnitude of the Casimir force between the plates was precisely as Casimir himself had predicted.

Spooky action–at–a–distance

Einstein was far from satisfied with the Copenhagen interpretation. Having sowed the seeds of the quantum revolution with his breathtakingly speculative paper in 1905, by the mid-1920s he was fast becoming quantum theory's most determined critic.

He was particularly concerned about the collapse of the wavefunction. If a single photon is supposed to be described by a wavefunction distributed over a region of space, where then is the photon supposed to *be* prior to the collapse? Before the act of measurement, the energy of the photon is in principle 'everywhere'. What then happens at the moment of the collapse? After the collapse, the energy of the photon is localized – it is 'here' and nowhere else. How does the photon get from being 'everywhere' to being 'here' instantaneously?

Einstein called it 'spooky action–at–a–distance'. He was convinced that this violated one of the key postulates of his own special theory of relativity: no object, signal or influence having physical consequences can travel faster than the speed of light.

Through the late 1920s and early 1930s, he challenged Bohr with a series of ever more ingenious 'thought experiments'. These were experiments carried out only in the mind as a way of exposing what Einstein believed to be quantum theory's fundamental flaws – its inconsistencies and incompleteness.

Bohr stood firm. He resisted the challenges, each time ably defending the Copenhagen interpretation and in one instance using Einstein's own general theory of relativity against him. But Bohr's case for the defence relied increasingly on arguments based on clumsiness, an essential and unavoidable disturbance of the system caused by the act of measurement, of the kind that he had criticized Heisenberg for. Einstein realized that he needed to find a challenge that did not depend directly on the kind of disturbance characteristic of a measurement, thus undermining Bohr's defence.

In 1935, together with two young theorists, Boris Podolsky and Nathan Rosen, Einstein devised the ultimate challenge. Imagine a physical system that produces two photons. We assume that the physics of the system constrains the two photons such that they are both produced in identical states of linear polarization.* We have no idea what these orientations are until we impose a reference frame by performing a measurement. According to the Copenhagen interpretation, until the measurement, the actual orientations are 'undetermined' – all orientations are possible in the wavefunction, just as the single photon in the two-slit experiment can be found anywhere on the photographic film prior to the collapse.

For the sake of clarity, we'll call these photons A and B. Photon A shoots off to the left, photon B to the right. We set up our polarizing film over on the left. We make a measurement and determine that photon A has vertical polarization along our laboratory z axis.

What does this mean for photon B? Obviously, we have not yet made any measurement on photon B, yet we can deduce that it, too, *must* have vertical polarization along this same axis. The physics of the process that produced the photons demands this. The polarization state of photon B appears to have suddenly changed, from 'undetermined' to vertical, even though we have made no measurement on it. And although we might in practice be constrained in terms of laboratory space, we could in principle wait for photon B to travel halfway across the universe before we make our measurement on A.

---

* This is a variation of the original Einstein–Podolsky–Rosen thought experiment, but it is entirely consistent with their approach.

All this talk of photons and their polarization states might seem rather esoteric, and it might be a little difficult to follow precisely what's going on. But it's important that we understand the nature of Einstein, Podolsky and Rosen's challenge and so fully appreciate what's at stake here.

Let's try the following (imperfect) analogy. Suppose I toss a coin. This happens to be a coin with some special properties. As it spins in the air, it splits into two coins. The operation of a 'law of conservation of coin faces' means that if the coin that lands over on the left (coin A) gives 'heads', then the coin that lands over on the right (coin B) gives 'heads' too. If coin A gives 'tails', then coin B also gives 'tails'. There are no circumstances under which we would expect to observe the results 'heads'–'tails' or 'tails'–'heads'.

Now suppose I toss the coin and we look to see what result we got for coin A. We see that it lands 'heads'. We know that coin B must also give the result 'heads' – the law of conservation of coin faces demands it.

Our instinct is to assume that the properties of the two coins are established at the instant they split from the original coin. The coins split apart and separate in mid-air, in some way that ensures that we eventually get correlated results – 'heads'–'heads' or 'tails'–'tails'. But assuming this implies that there are other variables involved of which we are ignorant, and that if this is all described by quantum theory, then the theory is in some sense incomplete.

In the photon example, the two particles are said to be 'entangled'. Because of the way they are produced, both photons are described by a single wavefunction. When we make a measurement on photon A, the wavefunction collapses and the polarization properties of both photons mysteriously become 'real'. If photon B is halfway across the universe, then the collapse must reach out across this distance. Instantaneously.

Likewise, quantum theory insists that our two coins don't possess fixed properties the instant they're created. It's almost as though they each oscillate back and forth from 'heads' to 'tails' as they spin to the ground. But the moment we see that coin A has landed 'heads', then coin B must also land 'heads', even though the two coins may fall far apart from each other.

Einstein, Podolsky and Rosen wrote: 'No reasonable definition of reality could be expected to permit this.'[12]

Despite what quantum theory says, Einstein, Podolsky and Rosen argued that it is surely reasonable to assume that when we make a measurement on photon A, this can in no way disturb photon B. What we choose to do with photon A cannot affect the properties and behaviour of B and hence the outcome of any subsequent measurement we might make on it. Under this assumption, we have no explanation for the sudden change in the polarization state of photon B, from 'undetermined' to vertical.

We conclude that there is, in fact, no change at all. Photon B must have vertical polarization all along. As there is nothing in quantum theory that tells us how the polarization states of the photons are determined at the moment they are produced, Einstein, Podolsky and Rosen concluded that the theory is incomplete.

'This onslaught came down upon us as a bolt from the blue,' wrote Belgian theorist Léon Rosenfeld, who was working with Bohr at his institute when news of this latest challenge reached Copenhagen.[13] The eminent English theorist Paul Dirac declared: 'Now we have to start all over again, because Einstein proved that it does not work.'[14]

Bohr's response was simply to restate the Copenhagen interpretation. He argued that we simply cannot get past the wave shadows and the particle shadows. Irrespective of the apparent puzzles caused by the need to invoke a collapse of the wavefunction, we just have to accept that that's the way it is. We have to deal with what we can measure and perceive. And these things are determined by the way we set up our experiment.

Bohr argued that it does not matter that the polarization state of photon B can be inferred from measurements we make on photon A. By setting up an experiment to measure the polarization of photon A with certainty (along the z axis, say), we deny ourselves the possibility of measuring the polarization along any other axis (x or y). And if we cannot exercise a choice, then the actual properties and behaviour of photon B are really rather moot. Even though there is no mechanical disturbance of photon B (no clumsiness), its properties and behaviour are nevertheless *defined* by the way we have set up the measurement on photon A.

The Einstein–Podolsky–Rosen thought experiment pushed Bohr to drop the clumsiness defence, just as Einstein had intended. But this left him with no alternative but to argue for a position that may, if anything, seem even more 'spooky'. The idea that the properties and behaviour of a quantum particle could somehow be influenced by how we *choose* to set up an apparatus an arbitrarily long distance away is very discomforting. Many years later the English physicist Anthony Leggett summarized this as follows:

> But in physics we are normally accustomed to require some positive reason before we accept a particular part of the environment as relevant to the outcome of an experiment. Now the [distant] polarizer … is nothing more than (e.g.) a calcite crystal, and nothing in our experience of physics indicates that the orientation of distant calcite crystals is either more or less likely to affect the outcome of an experiment than, say, the position of the keys in the experimenter's pocket or the time shown by the clock on the wall.[15]

For whom the Bell tolls

Physicists either accepted Bohr's arguments or didn't much care either way. Quantum theory was proving to be a very powerful structure, and any concerns about what it implied for our interpretation of reality were pushed to the back burner. The debate became less intense, although Einstein remained stubbornly unconvinced.

But Irish theorist John Bell continued to feel uncomfortable. Any attempt to eliminate the spooky action-at-a-distance implied in the Einstein, Podolsky and Rosen thought experiment involved the introduction of so-called 'hidden variables'. These are hypothetical properties of a quantum system that by definition are not accessible to experiment (that's why they're 'hidden') but which nevertheless govern those properties that we can measure. If, in the Einstein–Podolsky–Rosen experiment, hidden variables of some kind controlled the polarization states of the two photons such that they are fixed at the moment the photons are produced, then there would be no need to invoke the collapse of the wavefunction.* There would be no instantaneous change, no spooky action-at-a-distance.

Bell realized that if such hidden variables were assumed to exist, then in certain kinds of Einstein–Podolsky–Rosen-type experiments the hidden variable theory would predict results that disagreed with the predictions of quantum theory. It didn't matter that we couldn't be specific about precisely what these hidden variables were supposed to be. Assuming hidden variables of any kind means that the two photons are imagined to be *locally real* – they move apart as independent entities and continue as independent entities until one, the other or both are detected.

Going back to our coin analogy, a hidden variables extension would have the properties of the two coins fixed at the moment they split apart and separate. The coins are assumed to be locally real.

This seems perfectly reasonable, but quantum theory, in contrast, demands that the two photons or the two coins are non-local and entangled; they are described by a single wavefunction. They continue to be non-local and entangled until one, the other or both are detected, at which point the wavefunction collapses and the two photons or the two coins become localized, replete with the properties we measure.

This is Bell's theorem: 'If the [hidden variable] extension is local it will not agree with quantum mechanics, and if it agrees with quantum mechanics it will not be local. This is what the theorem says.'[16]

Bell was able to devise a relatively simple direct test. Hidden variable theories that establish some form of local reality predict experimental results that conform to something called Bell's inequality. Quantum theory does not.

Bell published his ideas in 1966. The timing was fortuitous. Sophisticated laser technology, optical instruments and sensitive detection devices were just becoming available. Within a few years the first practical experiments designed to test Bell's inequality were being carried out.

---

* Strictly speaking, it's not necessary to fix the polarization states at the moment the photons are produced. It's enough that the hidden variables are so fixed and determine how the photons interact with the polarizing film, such that if photon A is measured to be vertically polarized, photon B is, too.

The most widely known of these experiments were performed by French physicist Alain Aspect and his colleagues in the early 1980s. These made use of two high-powered lasers to produce excited calcium atoms, formed in an atomic 'beam' by passing gaseous calcium from a high temperature oven through a tiny hole into a vacuum chamber. Calcium atoms excited in this way undergo a 'cascade' emission, producing two photons in quick succession. The physics of the atom demands that angular momentum is conserved in this process, and the two photons are emitted in opposite states of circular polarization. This means that when they are passed through linear polarization filters, both photons will have the same linear polarization state, either both vertical or both horizontal.* The photons are entangled.

The two photons have different energies, and hence different frequencies (and different colours). The physicists monitored green photons (which we will call photons A) on the left and blue photons (photons B) on the right. Each polarizing filter was mounted on a platform which allowed it to be rotated about its optical axis. Experiments could therefore be performed for different relative orientations of the two filters, which were placed about 13 metres apart.

Imposing this separation distance meant that any kind of 'spooky' signal passing between the photons at the moment the wavefunction collapsed, 'informing' photon B of the fate of photon A, for example, would need to travel at about twice the speed of light.

The results came down firmly in favour of quantum theory. The physicists performed four sets of measurements with four different orientations of the two polarizing filters. This allowed them to test a generalized form of Bell's inequality. For the specific combination of orientations of the polarizing filters chosen, the generalized form of the inequality demands a value that cannot be greater than 2. Quantum theory predicts a value of 2.828.** The physicists obtained the result 2.697 ± 0.015.[17] In other words, the experimental result exceeded the

---

* Here 'polarizing films' is a shorthand for what was a complex bit of technical kit, including polarization analysers, photomultipliers, detectors and timing electronics that could identify when the photons belonged to the same pair.

** Actually, quantum theory predicts the value $2\sqrt{2}$.

limit predicted by Bell's inequality by almost fifty times the experimental error, a powerful, statistically significant violation.

In subsequent experiments the physicists modified their arrangement to include devices which could switch the paths of the photons, directing each of them towards two differently orientated polarizing filters. This prevented the photons from 'knowing' in advance along which path they would be travelling, and hence through which filter they would eventually pass. This was equivalent to changing the relative orientations of the two polarizing filters while the photons were in flight.

The physicists obtained the result 2.404 ± 0.080, once again in clear violation of the generalized form of Bell's inequality.

Similar experiments were carried out in August 1998 by a research group from the University of Geneva. They measured pairs of entangled photons using detectors positioned in Bellevue and Bernex, two small Swiss villages outside Geneva almost 11 kilometres apart. The results showed a clear violation of Bell's inequality. This suggests that any spooky action-at-a-distance would need to propagate from one detector to another with a speed at least twenty thousand times that of light.

Quantum entanglement has opened up intriguing possibilities in quantum information processing, cryptography and teleportation ('Beam me up, Scotty' for photons). Such possibilities are based inherently on the kind of non-local spookiness required to breach Bell's inequality and which Einstein had hoped to avoid. In May 2012, a team of physicists from various institutes in Austria, Canada, Germany and Norway, led by Austrian Anton Zeilinger, reported successful teleportation of photons from La Palma in the Canary Islands to Tenerife, 143 kilometres distant.[18]

There is no escaping the conclusion. Reality at the quantum level is decidedly non-local.

Testing non-local hidden variable theories

But the reality advocated by the proponents of hidden variable theories does not have to be a local reality. The influences of the hidden variables could be non-local. This would still leave us with an action-at-a-distance that is somewhat spooky, but at least it would get rid of the collapse of the wavefunction and the inherent quantum 'chanciness' that this implies.

Like Bell, Anthony Leggett was also rather distrustful of the Copenhagen interpretation of quantum theory. He understood that local hidden variable theories (of the kind that Bell had considered) are constrained by two important assumptions. In the first, we assume that whatever *result* we get for photon A, this can in no way affect the result of any simultaneous or subsequent measurement on the distant photon B, and vice versa.

The second assumption is rather subtle. We assume that however we *set up* the apparatus to make the measurement on photon A, this can in no way affect the result we get for photon B, and vice versa. Remember, in response to the challenge posed by Einstein, Podolsky and Rosen, Bohr argued that the properties and behaviour of photon B are *defined* by the way we set up the measurement on photon A.

We know from the experimental tests of Bell's inequality that one or other or both of these assumptions must be wrong and that something has to give. But the experiments do not tell which of them is invalid. Leggett wondered what would happen if, instead of abandoning both assumptions, we keep the 'result' assumption but relax the 'set-up' assumption.

In essence, relaxing the set-up assumption means that the behaviour of the photons and the results of measurements *can* be influenced by the way we set up our measuring devices, just as Bohr had argued. This is still pretty weird. It requires some kind of curious, unspecified non-local influence to be exerted by the choices we make in a possibly very distant laboratory.

In the context of our coin analogy, keeping the result assumption means that the result we get for coin B *cannot* depend on the result we get for coin A. We're still assuming that the faces of both coins are fixed at the moment they split apart. However, relaxing the set–up assumption means that the result we get for coin B can be influenced by *how we look* at coin A to see what result we got.

We can be reasonably confident that Einstein wouldn't have liked it.

By keeping the result assumption, Leggett defined a class of what he called 'crypto' non-local hidden variable theories. The most important thing to note about this class of theories is that the individual quantum particles are assumed to possess defined properties before we measure them. What we actually measure will, of course, depend on the way we set up our measuring devices, and changing these will affect the properties and behaviour of distant particles.

Here's the bottom line. This comes down to the rather simple question of whether or not quantum particles have the properties we assign to them *before the act of measurement*.

Leggett found that keeping the result assumption but relaxing the set-up assumption is still insufficient to reproduce all the predictions of quantum theory. Just as Bell had done in 1966, Leggett now derived a relatively simple inequality that could provide a direct test.

Experiments designed to test Leggett's inequality were performed in 2006 by physicists at the University of Vienna and the Institute for Quantum Optics and Quantum Information. The greatest difference between the predictions of quantum theory and the prediction of this whole class of crypto non-local theories arises for a specific arrangement of the polarizing filters.* For this arrangement, the class of non-local hidden variable theories predict a value for the Leggett inequality of 3.779. Quantum theory predicts 3.879, a difference of less than 3 per cent.

Nevertheless, the results were once again unequivocal. For the arrangement mentioned above, the experimental value was found to be 3.852±0.023, a violation of the Leggett inequality by more than three times the experimental error.[19]

Things-as-they-are-measured

In his response to the challenge from Einstein, Podolsky and Rosen, Bohr appeared to accept that there could be no 'mechanical disturbance', no 'ripple' effect arising from the outcome of the measurement on photon A. It's not clear from his writings if he believed that, despite the absence of a mechanical disturbance, both the result and set-up assumptions inherent in the presumption of local reality should be abandoned: '... even at this stage there is essentially the question of *an influence on the very conditions which define the possible types of predictions regarding the future behaviour of the system*'.[20]

---

* Again, I'm paraphrasing. The experiments were a lot more complicated than this.

However, the experimental tests of Leggett's inequality demonstrate that we must indeed abandon both the result *and* the set-up assumptions. The properties and behaviour of the distant photon B *are* affected by both the setting we use to measure photon A and the result of that measurement. It seems that no matter how hard we try, we cannot avoid the collapse of the wavefunction.

What does this mean?

It means that in experimental quantum mechanics we have run right up against what was previously perceived to be a purely philosophical barrier. The experiments are telling us that we can know nothing of reality-in-itself.

We have to accept that the properties we ascribe to quantum particles like photons, such as energy, frequency, spin, polarization, position ('here' or 'there'), are properties that have no meaning except in relation to a measuring device that allows them to be projected into our empirical reality of experience. We can no longer assume that the properties we measure necessarily reflect or represent the properties of the particles as they really are.

Perhaps even more disturbing is the conclusion that when we try to push further and ascribe to reality-in-itself properties that might help us to reconcile and understand our observations, we get it demonstrably wrong.

This is all strangely reminiscent of a famous philosophical conundrum. If a tree falls in the forest and there's nobody around to hear, does it make a sound?

Philosophers have been teasing our intellects with such questions for centuries. Of course, the answer depends on how we choose to interpret the use of the word 'sound'. If by sound we mean compressions and rarefactions in the air which result from the physical disturbances caused by the falling tree and which propagate through the air with audio frequencies, then we might not hesitate to answer in the affirmative.

Here the word 'sound' is used to describe a physical phenomenon – the wave disturbance carried by the air. But sound is also a human experience, the result of physical signals delivered by human sense organs which are synthesized in the mind as a form of perception.

As we have seen, sense perceptions can be described using chemical and physical principles, up until the point at which the perception becomes a mental experience. And the precise details of this process remain, at present, unfathomable.

The experiences of sound, colour, taste, smell and touch are all secondary qualities which exist only in our minds. We have no basis for our common-sense assumption that these secondary qualities reflect or represent reality as it really is. So, if we interpret the word 'sound' to mean a human experience rather than a physical phenomenon, then when there is nobody around, there is a sense in which the falling tree makes no sound at all.

What the experimental tests of Bell's and Leggett's inequalities tell us is much the same. We have no basis for our common-sense assumption that the properties of quantum particles such as photons reflect or represent reality as it really is.

Who would have thought that the theory of light would lead to such philosophy? It should now be apparent why the Reality Principle given in Chapter 1 is so structured.

But look back through this chapter. The conclusions of quantum theory may be utterly bizarre, but this is a theory founded on solid observational and experimental fact. It has been tested over and over again. Whether we like it or not, it is here to stay. It is 'true'. It describes the properties and behaviour of light better than any theory that has gone before, and is an essential component of the authorized version of empirical reality.

The Construction of Mass

Matter, Force and the
Standard Model of Particle Physics

A theory is the more impressive the greater the simplicity of its premises, the more different kinds of things it relates, and the more extended its area of applicability.

Albert Einstein[1]

Perhaps we would be more comfortable if the behaviour described in Chapter 2, which poses such bizarre philosophical conundrums, was in some way restricted to photons, those ghostly white ambassadors of morning. Alas, when in 1923 de Broglie speculated that there might be a connection between the wavelength of a quantum wave particle and its momentum, he was thinking not of photons, but of *electrons*. And, I can now admit, the two–slit interference pattern shown in Figure 1 on page 37 was produced not with a faint beam of light, admitting one photon after another, but with a faint beam of electrons.

It comes as something of a shock to realize that pictures such as Figure 1 relate to particles that we're more inclined to think of as tiny, but solid, bits of material substance. We then start to ask some really uncomfortable questions. If we thought it was spooky to lose sight of massless particles as they make their way (as waves) through a two–slit interference apparatus, then doing the same with electrons is surely downright embarrassing. As electrons pass – one after the other – through the apparatus, to be 'constituted' only when the wavefunction collapses, we have to ask ourselves: *what happens to the electrons' mass*?

Okay, okay. It's important to stay calm. Electrons are particles with mass, but this mass is nevertheless very, very small. The mass of an electron is about 0.9 thousandths of a billionth of a billionth of a billionth (9×10^{-31}) of a kilogram. If we lost an electron, I guess we would hardly miss it. Perhaps we can still get away with the idea that, because of their small size, electrons are susceptible to phantom-like, non-local behaviour of the kind we associate with photons. Larger particles or more massive structures should surely be less susceptible.

But this won't do. Quantum wave interference effects have been demonstrated with large molecules containing 60 and 70 carbon *atoms*. Superconducting quantum interference devices (SQUIDs, for short) have been used to demonstrate interference in objects of millimetre dimensions. These are dimensions you can *see*. These experiments involved combining SQUID states in which a billion electrons move clockwise around a small superconducting ring and another billion electrons move anticlockwise around the ring. In such a quantum superposition, in what direction are the electrons actually supposed to be moving?

It gets worse. In the standard model of particle physics, we learn that the property of mass of all the elementary particles – the particles that make up everything we are and everything we experience – is not an intrinsic or primary property of the stuff of material substance. It results from the interaction of quantum particles that would otherwise be massless with a mysterious energy field called the Higgs field which pervades the entire universe, like a modern-day ether. These interactions slow down the particles that interact with it, to an extent determined by the magnitude of their coupling to the field. We interpret this slowing down as inertia. And, ever since Galileo, we interpret inertia as a property of objects possessing mass.

We are forced to conclude that this interaction with the Higgs field, this slowing down, is actually what mass *is*.

We'd better take a closer look.

The forces of nature

When Einstein developed his theories of relativity and challenged Bohr over the interpretation of quantum theory in the 1920s, it was believed that there were just two forces of nature – electromagnetism and gravity. Early attempts to construct a unified theory capable in principle of describing all the elementary particles and their interactions therefore involved reconciling just these two forces in a single framework.

But two forces of nature aren't enough to account for the properties of atoms as these came to be understood in the early 1930s.

In 1932, English physicist James Chadwick discovered the neutron, an electrically neutral particle which, together with the positively charged proton, forms the building blocks of all atomic nuclei. It was now understood that each chemical element listed in the periodic table consists of atoms. Each atom consists of a nucleus composed of varying numbers of protons and neutrons. Each element is characterized by the number of protons in the nuclei of its atoms. Hydrogen has one, helium two, lithium three, and so on to uranium, which has 92. It is possible to create elements heavier than uranium in a particle accelerator or a nuclear reactor, but they do not occur in nature.

This was all very well, but it posed something of a dilemma. We know that like charges repel each other, so how could all those positively charged protons be squeezed together and packed so tightly inside an atomic nucleus, and yet remain stable? Careful experimental studies revealed that the strengths of the interactions between protons and protons inside the nucleus are very similar in magnitude to those between protons and neutrons. None of this made any sense, unless the force governing these interactions is very different from electromagnetism. And very much stronger, able to overcome the force of electrostatic repulsion threatening to tear the nucleus apart.

This suggested the existence of another force, which became known as the *strong nuclear force*, binding protons and neutrons together in atomic nuclei.

This was not quite the end of the story. It had been known since the late nineteenth century that certain isotopes of certain elements – atoms with the same numbers of protons in their nuclei but different numbers of neutrons – are unstable. For example, the isotope caesium-137 contains 55 protons and 82 neutrons. It is radioactive, with a half-life (the time taken for half the radioactive caesium-137 to disintegrate) of about thirty years. The caesium-137 nuclei disintegrate spontaneously through one or more nuclear reactions.

There are different kinds of radioactivity. One kind, which was called beta-radioactivity by New Zealand physicist Ernest Rutherford in 1899, involves the transformation of a neutron in a nucleus into a proton, accompanied by the ejection of a high-speed electron (also known as a 'beta particle'). This is a natural form of alchemy: changing the number of protons in the nucleus changes its chemical identity. In fact, caesium-137 decays by emission of a high-speed electron to produce barium-137, which contains 56 protons and 81 neutrons. This implies that the neutron is an unstable, composite particle, and so not really elementary at all. Left to its own devices, an isolated neutron will decay spontaneously in about 15 minutes.

The origin of beta-radioactivity was a bit of a puzzle. What was an *electron* supposed to be doing inside the nucleus? But it posed an even bigger problem. It could be expected that in this process of radioactive decay, energy and momentum should be conserved, just as they are conserved in every chemical and physical change that has ever been studied. But a careful accounting of the energies and momenta involved in beta-radioactivity showed that the numbers just didn't add up. The theoretical energy and momentum released by the transformation of a neutron inside the nucleus could not all be accounted for by the energy and momentum of the ejected electron.

In 1930, Wolfgang Pauli reluctantly suggested that the energy and momentum 'missing' in the reaction were being carried away by an as yet unobserved, light, electrically neutral particle which interacts with virtually nothing. It came to be called a *neutrino* (Italian for 'small neutral one'). At the time it was judged that it would be impossible to detect, but it was first discovered experimentally in 1956.

The strong nuclear force was judged to be about a hundred times stronger than electromagnetism. But the force governing beta-radioactive decay was found to be much weaker, about ten billionths of the strength of the electromagnetic force. It was also clear that this weak force acted on nuclear particles – protons and neutrons – although electrons, too, were somehow involved. There was no choice but to conclude that this was evidence for yet another force of nature, which became known as the *weak nuclear force*.

Instead of two fundamental forces there were now four: the strong and weak nuclear forces, electromagnetism and gravity. Constructing a quantum theory that could accommodate three of these forces (the exception being gravity) took another forty years. It is called the standard model of particle physics, and it is one of the most successful theories of physics ever devised.

Spinning electrons and anti-matter

Two versions of quantum theory were developed in the 1920s and were applied principally to the study of the properties and behaviour of electrons. These were *matrix mechanics*, devised in 1925 by Heisenberg, and *wave mechanics*, devised in late 1925/early 1926 by Austrian physicist Erwin Schrödinger. Although these were rival theories, Schrödinger himself demonstrated that they are, in fact, equivalent; one and the same theory expressed in two different mathematical 'languages'.

These early quantum theories were further developed and elaborated, and cast into other mathematical forms that have since proved to be useful (or, at least, they avoid some of the metaphysical baggage that the early theories tended to carry). Quantum theory is still very much a theory of contemporary physics.

But the equations of quantum theory describe individual quantum particles or systems. They describe how the energies associated with various motions of the particles or systems are 'quantized', able to admit or remove energy only in discrete lumps, or quanta. Adding quanta therefore involves increasing the energy, promoting the particles or systems to higher-energy quantum states. Taking quanta away likewise involves removing energy, often in the form of emitted photons, demoting the particles or systems to lower-energy states.

To describe correctly such 'dynamics' of quantum systems, the theory must in principle conform to the stringent requirements of Einstein's special theory of relativity. This ensures that the laws of nature that we can observe or measure are guaranteed to be the same independently of how fast the observer or measuring device might happen to be moving in relation to the object under study.* It also

---

* At first sight this might seem an odd thing to demand, but it ensures that the 'laws' of nature that we deal with in our theories are genuine laws, independent of any specific point of view.

ensures that the speed of light retains its privileged status as an ultimate speed that cannot be exceeded.*

Einstein himself identified one important consequence of the special theory. This was a deep connection between energy and mass reflected in the world's most famous scientific equation, $E = mc^2$, or energy equals mass multiplied by the speed of light squared. This is interpreted to mean that mass is a form of energy, and from energy can spring mass.

Now this might already cause us to pause for a moment's quiet reflection on what it may mean for our interpretation of mass. I confess that, as a young student, the notion that mass represents a vast reservoir of energy somehow made it seem even more tangible, and 'solid'. It didn't shake my naïve conviction that mass must surely be an intrinsic property or primary quality of material substance.

In essence, producing a so-called 'relativistic' theory – one that meets the requirements of the special theory of relativity – is all about ensuring that the theory treats *time* as a kind of fourth dimension, on an equal footing with the three dimensions of space. In late 1927, Paul Dirac extended the early version of quantum theory and made it conform to the special theory of relativity.

The resulting theory predicted that the electron should possess a spin quantum number s equal to ½ and two different spin orientations, labelled spin-up and spin-down.[2] This was something of a revelation. That electrons possess an intrinsic spin angular momentum had been shown experimentally some years previously, but neither matrix nor wave mechanics had predicted this behaviour.

Dirac was a mathematician first, a physicist second. In devising his theoretical equation he had followed his mathematical instincts, and when these seemed to suggest some relatively unphysical conclusions,

---

* On 23 September 2011, physicists at CERN's OPERA experiment reported results that suggested that neutrinos travelling the 730 kilometres from Geneva to Italy's Gran Sasso laboratory do so with a speed slightly greater than that of light. Data collected from over 16,000 events recorded over a three-year period suggested that the particles were arriving about 61 billionths of a second earlier than expected. Such faster-than-light neutrinos would have represented a fundamental unravelling of Einstein's special theory of relativity and thus the current authorizsed version of reality. Many scientists were sceptical of the results and some argued that they couldn't be correct. On 22 February 2012, the OPERA results were shown to be erroneous, and a loose fibre-optic cable was blamed. The claims were withdrawn, and a couple of high-profile members of the collaboration resigned their positions.

he had nevertheless stood firm.[3] The theory produced twice as many solutions as he thought he had needed. Two of these solutions correspond to the spin-up and spin-down orientations of the electron, each carrying positive energy. But there were another two solutions of 'negative' energy. Although these solutions were hard to understand, they couldn't be ignored.

In 1931, Dirac finally conceded that the two negative-energy solutions actually correspond to positive-energy spin-up and spin-down orientations of a *positively* charged electron. He had discovered the existence of antimatter, a previously unsuspected form of material substance. There had been no hints of antimatter in any experiment performed to that time and no good reason to suspect that it existed. It was completely 'off the wall'.

Yet within a year, the positive electron had been discovered in experiments on cosmic rays. It was named the *positron*. Dirac's mathematical instincts were proved right.

Quantum fields and second quantization

Dirac's theory is actually a kind of *quantum field theory*. A field in physics is defined in terms of the magnitude of some physical property distributed over every point in space and time. Sprinkle iron filings on a sheet of paper held above a bar magnet. The iron filings organize themselves along the 'lines of force' of the magnetic field, reflecting the strength of the field and its direction, stretching from north to south poles. The field exists in the 'empty' space around the outside of the bar of magnetic material.

The nature of the field depends on the nature of the property being measured. The field may be *scalar*, meaning that it has magnitude but acts in no particular direction in space – it points but it doesn't push or pull. It may be *vector*, with both magnitude and direction, like a magnetic field or a Newtonian gravitational field. Finally, it may be *tensor*, a more complicated version of a vector field for situations in which the geometry of the field requires more parameters than would be required for three 'Cartesian' x, y and z directions.

The field described by Dirac's theory is in fact none of the above. It is a *spinor field*. Spinors were discovered by mathematician Elié Cartan in 1913. They are vectors, but not in the sense of vectors in 'ordinary' space, and they cannot be constructed from ordinary vectors. They are, of course, spin vectors, related to the spin orientations of the electron. To make his theory conform to the requirements of special relativity, Dirac needed a spinor field with four components – two for the electron and two for the positron.

It might be best to think of the field used in Dirac's theory as an 'electron field', the quantum field for which the corresponding quantum particle is the electron. The particle is, in essence, a fundamental field quantum, a basic fluctuation or disturbance of the field.

In contrast to conventional quantum theory, when we add or remove quanta in a quantum field theory, we're adding or removing particles to or from the field itself. We add energy to the field in the form of particles, rather than adding it to individual quantum particles. This is sometimes known as 'second quantization'.

The strange theory of light and matter

Dirac's theory represented a major advance, but as a field theory it seemed rather obscure. It wasn't obvious from the structure of his equation that it met the demands of special relativity (though it did); neither was it clear how the theory should be related to much better understood classical field theories, such as Maxwell's theory of electromagnetism.

The attentions of some physicists were drawn instead to an alternative approach. Why not start with a well-understood classical wave field, and then find a way to impose quantum conditions on it? The obvious place to start was the field described by Maxwell's equations. If it could be done, the resulting quantum field theory would describe the interactions between an electron field (whose quanta are electrons) and the electromagnetic field (whose quanta are photons).

It began to dawn on physicists working on early versions of quantum field theory that they had figured out a very different way to understand how forces between particles actually work. Let's imagine that two electrons are 'bounced' off each other (the technical term is 'scattered'). We can suppose that as the two electrons approach each other in space and in time, they feel the mutually repulsive force generated by their negative charges.

But how? We might speculate that each moving electron generates an electromagnetic field and the mutual repulsion is felt in the space where these two fields overlap, much like we feel the repulsion between the north poles of two bar magnets in the space between them as we try to push them together. But in the quantum domain, fields are also associated with particles, and interacting fields with interacting particles. In 1932, German physicist Hans Bethe and Italian Enrico Fermi suggested that this experience of force is the result of the *exchange of a photon* between the two electrons.

As the two electrons come closer together, they reach some critical distance and exchange a photon. In this way the particles experience the electromagnetic force. The exchanged photon carries momentum from one electron to the other, thereby changing the momentum of both. The result is recoil, with both electrons changing speed and direction and moving apart.

The exchanged photon is a 'virtual' photon, because it is transmitted directly between the two electrons and we don't actually see it pass from one to the other. In fact, there's no telling in which direction the photon actually goes. In diagrams drawn to represent the interaction, the passage of a virtual photon is simply illustrated using a squiggly line, with no direction indicated between the electrons.

Here was another revelation. The photon was no longer simply the quantum particle of light. It had become the 'carrier' of the electromagnetic force.

Heisenberg and Pauli had tried to develop a formal quantum field theory of electromagnetism a few years before, in 1929. But this theory was plagued with problems. The worst of these was associated with the 'self-energy' of the electron. When an electric charge moves through space, it generates an electromagnetic field. The 'self-energy' of the electron results when an electron interacts with its own self-generated electromagnetic field. This interaction caused the equations to 'blow up', producing physically unrealistic results. Some terms in the equation mushroomed to infinity.

A solution to this problem would be found only in 1947, when physicists were able to return to academic science after spending the war years working on the world's first atomic weapons.

Suppose the self-energy of the electron appears (using $E = mc^2$) as an additional contribution to the electron's mass. The mass that we observe in experiments would then include this additional contribution. It would be equal to an intrinsic or 'bare' mass plus an 'electromagnetic' mass arising from the electron's interaction with its own electromagnetic field.

The 'bare' mass is a purely theoretical quantity. It is the mass that the electron would possess if it could ever be isolated from its own electromagnetic field. The mass that we have to deal with is the observed mass, so the theory has to be rewritten in terms of this. In other words, the theory has to be 'renormalized'.

Suppose further that we want to apply quantum field theory to the calculation of the energies of the quantum states of an electron in a hydrogen atom. The theory predicts an infinite self-energy associated with the electron interacting with its own electromagnetic field. We identify this as an infinite contribution to the electron mass. But the equations for a freely moving electron also contain the same infinite mass contribution. So, what if we now subtract the expression for a free electron from the expression for the electron in a hydrogen atom? Would the two infinite terms cancel to give a finite answer?

It sounds as though subtracting infinity from infinity should lead only to nonsense, but it worked. A formal theory of quantum electrodynamics (QED), essentially a quantum version of Maxwell's theory, was eventually devised by rival American physicists Richard Feynman and Julian Schwinger, and independently by Japanese theorist Sin-Itiro Tomonaga. Although the approaches adopted by Schwinger and Tomonaga were similar, Feynman's was distinctly different, relying on pictorial representations of the different kinds of possible interactions which came to be called 'Feynman diagrams'. English physicist Freeman Dyson subsequently demonstrated that their different approaches were entirely equivalent.

Not everyone was comfortable with the apparent sleight of hand involved in renormalization. Dyson asked Dirac: 'Well, Professor Dirac, what do you think of these new developments in quantum electrodynamics?' Dirac, the mathematical purist, was not enamoured: 'I might have thought that the new ideas were correct if they had not been so ugly.'[4]

Ugly or not, there was no denying the power of the resulting theory. The g-factor for the electron, a physical constant governing the strength of the interaction of an electron with an external magnetic field, is predicted by QED to have the value 2.00231930476. The comparable experimental value is 2.00231930482. 'This accuracy', wrote Feynman, 'is equivalent to measuring the distance from Los Angeles to New York, a distance of over 3,000 miles, to within the width of a human hair.'[5]

The particle zoo

Dirac once speculated that it might one day be possible to describe all matter in terms of just one kind of elementary particle, some kind of ultimate quantum of 'stuff'. Alas, as experimental physicists set to work with cosmic rays and early particle accelerators in the 1930s and 40s, Dirac's dream was shattered. Far from discovering an underlying simplicity in nature, they discovered instead a bewildering complexity.

American physicist Carl Anderson had discovered the positron in 1932. Four years later he discovered another particle, a heavier version of the electron, with a mass about 200 times that of an ordinary electron. This simply did not fit with any preconceptions of how the elementary building blocks of nature should be organized. Galician-born American physicist Isidor Rabi demanded to know: 'Who ordered that?'[6] The new particle was given several names, but is today called the *muon*.

In 1947, another new particle was discovered in cosmic rays by British physicist Cecil Powell. This was found to have a slightly larger mass than the muon; 273 times that of the electron. It came in positive, negative and, subsequently, neutral varieties. This was called the *pion*. As techniques for detecting particles became more sophisticated, the floodgates opened. The pion was quickly followed by the positive and negative kaon and the neutral lambda particle.

What was going on? One physicist expressed the prevailing sense of frustration: '… the finder of a new elementary particle used to be rewarded by a Nobel Prize, but such a discovery now ought to be punished by a $10,000 fine'.[7] Names proliferated. Responding to a question from one young physicist, Fermi remarked: 'Young man, if I could remember the names of these particles, I would have been a botanist.'[8]

In an attempt to give some sense of order to this 'zoo' of particles, physicists introduced a new taxonomy. They defined two principal classes: *hadrons* (from the Greek *hadros*, meaning thick or heavy) and *leptons* (from the Greek *leptos*, meaning small).

The class of hadrons includes a subclass of *baryons* (from the Greek *barys*, also meaning heavy). These are heavier particles which experience the strong nuclear force. The proton and neutron are baryons. It also includes the subclass of *mesons* (from the Greek *mésos*, meaning 'middle'). These particles experience the strong force but are of intermediate mass. Examples include pions and kaons.

The class of leptons includes the electron, muon and neutrino. These are light particles which do not experience the strong nuclear force.

The baryons and the leptons are *matter particles*. They are also all fermions, characterized by half-integral spins, which obey Pauli's exclusion principle.

In addition to these there are *force particles*. These include the photon. These are *bosons*, characterized by integral spins. They are not subject to Pauli's exclusion principle. Bosons with zero spin are also possible, but these are not force particles. Mesons are examples.

The taxonomy was a little like organizing the chemical elements into a periodic table. It helped to establish the patterns among the different particle classes but gave no real clue to the underlying explanation.

The left hand of the electron

Things were about to get even more complicated. A wavefunction, such as a sine wave, moves up and down as it oscillates between peak and trough. Parts of the wavefunction have positive amplitude (as it rises to the peak and falls back) and parts have negative amplitude (as it dips below the axis heading for the trough and comes back up again).

The *parity* of the wavefunction is determined by its behaviour as we change the signs of the spatial co-ordinates in which the wave propagates. Think of this as changing left for right or up for down or front for back. Changing the signs of all three co-ordinates simultaneously is then a bit like reflecting the wavefunction in a special kind of mirror that also inverts the image and its perspective. The image is inverted left-to-right and up-to-down, and the front goes to the back as the back is brought forward to the front.

If reflecting the wavefunction in such a 'parity mirror' doesn't change the sign of the wavefunction's amplitude, then the wavefunction is said to possess even parity. However, if the wavefunction amplitude does change sign (from positive to negative or negative to positive), then the wavefunction is said to have odd parity.

Parity, like spin, is a property without many analogies in classical physics that are not thoroughly misleading. It is closely connected with and governs angular momentum in elementary particle interactions. As far as the physicists could tell, in all electromagnetic and nuclear interactions, parity is something that is conserved, like angular momentum itself. In other words, if we start with particles which when combined together have even parity, then we would expect that the particles that result from some physical process would also combine to give even parity. Likewise for particles with overall odd parity.

This seemed consistent with the physicists' instincts. How could it be possible for the immutable laws of nature to favour such seemingly human conventions of left vs right, up vs down, front vs back? Surely no natural force could be expected to display such 'handedness'?

As reasonable as this seems, in fact it is not consistent with what we observe. Parity is conserved in all electromagnetic interactions and processes involving gravity and the strong nuclear force. But nature exhibits a peculiar 'handedness' in interactions involving the weak force.

The first definitive example of such parity violation came from a series of extremely careful experiments conducted towards the end of 1959 by Chien-Shiung Wu, Eric Ambler and their colleagues at the US National Bureau of Standards laboratories in Washington DC. These involved the measurement of the direction of emission of beta-electrons from atoms of radioactive cobalt-60, cooled to near absolute zero temperature, their nuclei aligned by application of a magnetic field. A symmetrical pattern of beta-electron emission would have suggested that no direction is specially favoured, and that parity is conserved. The asymmetrical pattern that was actually observed indicated that parity is not conserved.

The experiments were unequivocal, and similar results have been observed in many other weak force interactions. Parity is not conserved in processes governed by the weak nuclear force. In fact, by convention, only 'left-handed' particles and 'right-handed' anti-particles actually undergo weak force interactions.

Nobody really understands why.

Unifying the electromagnetic and weak nuclear forces

The ranges and strengths of electromagnetism and the weak nuclear force are so very different that it appears at first sight impossible to reconcile them. But what if, reasoned Schwinger in 1941, the carrier of the weak nuclear force is actually a massive particle equal in size to a couple of hundred times the proton mass? If a force is carried by such a massive particle then its range becomes very limited, as (unlike photons) massive particles are very sluggish. The force would also become considerably weaker.

Schwinger realized that if the mass of such a weak force carrier could be somehow 'switched off', then the weak force would have a range and strength similar in magnitude to electromagnetism. This was the first hint that it might be possible to *unify* the weak and electromagnetic forces into a single 'electro-weak' force.

The logic runs something like this. Despite the fact that they appear so very different, the electromagnetic and weak nuclear forces are in some strange way manifestations of the same 'electro-weak' force. They appear very different because something has happened to the carrier of what we now recognize as the weak force. Unlike the photon, it somehow gained a lot of mass, restricting the range of the force and greatly diminishing its strength relative to electromagnetism.

Now, the key question was this. What happened to the carrier of the weak force to make it so heavy?

The challenge was taken up in the 1950s by Schwinger's Harvard graduate student, Sheldon Glashow. After some false starts, Glashow developed a quantum field theory of electro-weak interactions in which the weak force is carried by *three* particles. Two of these particles – now called the W^+ and W^- – are necessary to account for the fact that, unlike the case of electromagnetism, electrical charge is transferred in weak force interactions (a neutral neutron decays into a positively charged proton, for example). In effect, these particles are electrically charged, heavy versions of the photon. A third, neutral force carrier is also demanded by the theory. This was subsequently called the Z^0.

In this scheme beta-radioactivity could be explained this way. A neutron emits a massive W^- particle and turns into a proton. The short-lived W^- particle then decays into a high-speed electron (the beta-particle) and what is now understood to be an anti-neutrino.

But there were more problems. The quantum field theory that Glashow developed predicted that the force carriers should be massless. And if the masses of the force carriers were added to the theory 'by hand', the equations couldn't be renormalized.

So, precisely how *did* the W^+, W^- and Z^0 particles gain their mass?

The 'God particle' and the origin of mass

The solutions to these puzzles were found in the seven-year period 1964–71. The answer to the mass question was to invoke something called *spontaneous symmetry-breaking*.

This is a rather grand phrase for what is a relatively simple phenomenon. There are many examples of spontaneous symmetry-breaking we can find in 'everyday' life. If we had enough patience, we could imagine that we could somehow balance a pencil finely on its tip. We would discover that this is a very symmetric, but very unstable, situation. The vertical pencil looks the same from all directions.

But tiny disturbances in our immediate environment (such as small currents of air) are enough to cause it to topple over. When this happens, the pencil topples over in a specific, though apparently random, direction. The horizontal pencil no longer looks the same from all directions, and the symmetry is said to be spontaneously broken.

We don't need a PhD to work out that the less symmetrical state with the pencil lying on the table, pointing in a specific direction, has a lower energy than the more symmetrical state with the pencil balanced on its tip. Physicists call this more stable state the *ground state* of the system.

They reserve the special term 'vacuum state' for the quantum state of lowest possible energy – the ground state with *everything* removed (the pencil, the table, me, you and every last electron and photon). Now, let's set aside for a moment everything we learned in the last chapter about quantum fluctuations in the vacuum and think of it as just 'empty' space, what the philosophers of Ancient Greece used to call 'void'. Of course, such an empty space would be highly symmetrical – like the pencil, it would look the same from all possible directions.

But aside from random quantum fluctuations, what if empty space isn't actually empty? What if it contains a quantum field that, like the air currents that tip the pencil, spontaneously breaks the symmetry, giving a state of even lower energy?

When applying this idea to a particular problem in the quantum field theory of superconducting materials, American physicist Yoichiro Nambu realized that spontaneous symmetry-breaking can result in the formation of particles with mass. Some years later he wrote:

> What would happen if a kind of superconducting material occupied all of the universe, and we were living in it? Since we cannot observe the true vacuum, the [lowest-energy] ground state of this medium would become the vacuum, in fact. Then even particles which were massless … in the true vacuum would acquire mass in the real world.[*]

Physicists call this lower-energy, more stable vacuum state a 'false' vacuum. False, because although it contains nothing of obvious substance, it isn't empty. It contains a quantum field that breaks the symmetry.

It was now possible to put two and two together, although the path to a formal solution was still rather tortuous. In 1964 there appeared a series of papers detailing a mechanism for spontaneous symmetry-breaking applied to quantum field theory. These were published independently by Belgian physicists Robert Brout and François Englert, English physicist Peter Higgs at Edinburgh University, and Gerald Guralnik, Carl Hagen and Tom Kibble at Imperial College in London. The mechanism is commonly referred to as the *Higgs mechanism*.

It works like this. Prior to breaking the symmetry, the electro-weak force is carried by four massless particles which, for the sake of simplicity, we will call the W^+, W^0, W^- and B^0. A massless field particle has two 'degrees of freedom' and moves at the speed of light. For the photon, these two degrees of freedom are related to the particle's spin orientations. We perceive these different spin states as left-circular and right-circular polarization or, when combined in the right way, vertical (up/down) and horizontal (left/right) polarization. Although space is three-dimensional, special relativity forbids the photon from having polarization in a third (forward/back) direction.

In a conventional quantum field theory of the kind that Glashow developed, there is nothing to change this situation. Massless particles continue to be massless.

But what if we now assume that the vacuum isn't actually empty? What happens if we introduce a false vacuum by adding a background quantum field (often called a Higgs field) to break the symmetry? In this situation, massless particles interact with the Higgs field and acquire a third degree of freedom. The W^+ and W^- particles acquire 'depth' and get 'fat'. This act of gaining three-dimensionality is like applying a brake: the particles slow down to an extent which depends on the strength of their interaction with the field. The field drags on them like molasses.

In other words, the interactions of each particle with the Higgs field are manifested as a resistance to the particle's acceleration.[*]

Now, we tend to think of an object's resistance to acceleration as the result of its inertial mass. Our instinct is to assume that mass is a primary or intrinsic quality, and we identify inertial mass with the amount of 'stuff' that the object possesses. The more stuff it has, the harder it is to accelerate.

But the Higgs mechanism turns this logic on its head. The extent to which an otherwise massless particle's acceleration is resisted by the Higgs field is now interpreted as the particle's (inertial) mass. Mass has suddenly become a secondary quality. It is the result of an interaction, rather than something that is intrinsic to matter.

The W^0 and B^0 particles of the electro-weak force mix together to produce the massive Z^0 particle and the massless photon. We associate the massive W^+, W^- and Z^0 particles with the (now broken) weak force and the massless photon with electromagnetism.

In their publications, Brout, Englert, Higgs, Guralnik, Hagen and Kibble had not sought to apply this mechanism to the problem of the carriers of the electro-weak force. This task fell to American physicist Steven Weinberg. Weinberg had been struggling to apply the Higgs mechanism to a quantum field theory of the strong nuclear force, when he was suddenly struck by another idea: 'At some point in the fall of 1967, I think while driving to my office at MIT, it occurred to me that I had been applying the right ideas to the wrong problem.'[10]

---

[*] Note that it is accelerated motion which is impeded. Particles moving at a constant velocity are not affected by the Higgs field. For this reason the Higgs field is not in conflict with the demands of Einstein's special theory of relativity.

'My God,' he exclaimed to himself, 'this is the answer to the weak interaction!'[11]

In November 1967, Weinberg published a paper detailing a unified electro-weak quantum field theory. In this theory spontaneous symmetry-breaking using the Higgs mechanism is responsible for the differences between electromagnetism and the weak nuclear force in terms of strength and range. These differences can be traced to the properties of the W^+, W^- and Z^0 particles, which gain mass, and the photon, which remains massless. Weinberg estimated that the W particles would each have a mass about 85 times that of the proton, and the Z^0 would be slightly heavier, with a mass about 96 times the proton mass.

A quantum field must have an associated field particle. In 1964, Higgs had referred to the possibility of the existence of what would later become known as a 'Higgs boson', the elementary particle of the Higgs field. Three years later, Weinberg had found it necessary to introduce a Higgs field with four components. Three of these give mass to the W^+, W^- and Z^0 particles. The fourth appears as a physical particle – a Higgs boson with a spin quantum number of 0. If the Higgs mechanism really is responsible for the masses of the W^+, W^- and Z^0 particles, then not only should these particles be found with the predicted masses, but the Higgs boson should be found, too.

In Britain, Tom Kibble introduced the idea of spontaneous symmetry-breaking to one of his colleagues at Imperial College, Pakistan-born theorist Abdus Salam. Salam independently developed a unified electro-weak theory at around the same time. Both Weinberg and Salam believed that the theory should be renormalizable, but neither was able to prove this.

The proof followed in 1971. By sheer coincidence, Dutch theorists Martinus Veltman and Gerard 't Hooft rediscovered the field theory that Weinberg had first developed four years earlier, but they could now show how it could be renormalized. 't Hooft had initially thought to apply the theory to the strong nuclear force, but when Veltman asked a colleague about other possible applications, he was pointed in the direction of Weinberg's 1967 paper. Veltman and 't Hooft now realized that they had developed a fully renormalizable quantum field theory of electro-weak interactions.

This was all fine in theory, but what of experiment?

The electro-weak theory makes three principal predictions. First, if the weak nuclear force really does require three force carriers, then the exchange of one of these – the Z^0 – should result in weak force interactions involving no change in charge. To all intents and purposes, these interactions look just like interactions involving the exchange of a photon. The physicists call such interactions 'weak neutral currents' – they involve the weak force and result in no exchange of electrical charge (they are neutral).

Such currents were identified in particle accelerator experiments performed at CERN in Geneva in 1973, and subsequently at the US National Accelerator Laboratory (which was renamed Fermilab in 1974).

Second, Weinberg had predicted the masses for all the weak force carriers. At the time he made these predictions there was no particle accelerator large enough to observe them. But in the years that followed, a new generation of particle colliders was constructed in America and at CERN. The discovery of the W particles at CERN was announced in January 1983, with masses 85 times that of the proton, just as Weinberg had predicted. The discovery of the Z^0 was announced in June that year, with a mass about 101 times that of a proton.[12]

The third prediction concerns the existence of the Higgs boson. Given that the Higgs mechanism allows the masses of the weak force carriers to be predicted with such confidence, the existence of a Higgs field – or something very like it – seems a 'sure thing'. However, there are alternative theories of symmetry-breaking that do not require a Higgs field, and there remain problems with the electro-weak theory which erode our confidence somewhat and suggest that we might not yet have the full story.

The question of whether or not the Higgs boson exists in nature is therefore of fundamental importance.

On 4 July 2012, scientists at CERN's Large Hadron Collider declared that they had discovered a new particle 'consistent' with the standard model Higgs boson. After hearing presentations from the two detector collaborations, ATLAS and CMS, CERN director-general Rolf Heuer declared: 'As a layman I would say that I think we have it. Do you agree?'[13]

The new boson was found to have a mass around 133 times that of a proton[14] and interacts with other standard model particles in precisely the way expected of the Higgs. Apart from some slight anomalies, notably an observed enhancement in the decay into two photons (H → γγ), the new boson's decay modes to other particles have the ratios expected of a standard model Higgs. Whilst the ATLAS and CMS experiments were clear that this is a boson, neither could be clear on the precise value of its spin quantum number, which on the basis of the experimental results could be 0 or 2. However, the only particle anticipated to have spin-2 is the graviton, the purported carrier of the force of gravity. Spin-0 is therefore much more likely.

Although further research is required to characterize the new particle fully, the default assumption is that this is indeed a Higgs boson. But *which* Higgs boson? The standard model needs just one to break the electro-weak symmetry, though there are theories that extend beyond the standard model which demand rather more. The only way to find out precisely what kind of particle has been discovered is to explore its properties and behaviour in further experiments.

CERN commented:

> Positive identification of the new particle's characteristics will take considerable time and data. But whatever form the Higgs particle takes, our knowledge of the fundamental structure of matter is about to take a major step forward.[15]

Although the Higgs mechanism was invoked to explain how the carriers of the weak force acquire their mass, it is now understood that this is the mechanism by which *all* elementary particles gain mass. In a book published in 1993, American physicist Leon Lederman emphasized* the fundamental role played by the Higgs boson and called it the *God particle*.

Not many practising physicists like it, but it is a name that has stuck.

Three (actually, six) quarks for Muster Mark!

The flurry of experimental activity that established the 'zoo' of particles in the late 1950s demanded some simplifying theoretical scheme.

---

* Or over-emphasized, depending on your point of view.

Significant advances were made in the early 1960s. Further patterns identified by American theorist Murray Gell-Mann and Israeli Yuval Ne'eman called attention to the possibility that the hadrons might actually all be composite particles made of three even more elementary particles then unknown to experimental science.

The patterns suggested that hadrons such as the proton and neutron should no longer be considered to be elementary particles, but are instead composed of smaller constituents. But there was a big problem.

When Gell-Mann's colleague Robert Serber broached this idea over lunch at Columbia University in New York in 1963, Gell-Mann was initially dismissive.

> It was a crazy idea. I grabbed the back of a napkin and did the necessary calculations to show that to do this would mean that the particles would have to have fractional electric charges – $-\frac{1}{3}$, $+\frac{2}{3}$, like so – in order to add up to a proton or neutron with a charge of plus or zero.[16]

By this time, more than fifty years had elapsed since the idea of a fundamental unit of electrical charge had been established, and there had been no hints that there might be exceptions. But despite these worrying implications, there was no doubting that a system of smaller elementary particles did provide a potentially powerful explanation for the pattern of hadrons. Gell-Mann called these odd new particles 'quarks'.*

At that time the pattern of particles demanded three quarks, which were called 'up', with a charge of $+\frac{2}{3}$, 'down', with a charge of $-\frac{1}{3}$, and 'strange', a heavier version of the down quark, also with a charge of $-\frac{1}{3}$. The baryons known at that time could then be formed from various permutations of these three quarks and the mesons from combinations of quarks and anti-quarks.

In this scheme the proton consists of two up quarks and a down quark, with a total charge of $+1$. The neutron consists of an up quark

---

* At around the same time, American physicist George Zweig developed an entirely equivalent scheme based on a fundamental triplet of particles that he called 'aces'. Zweig struggled to get his papers published, but Gell-Mann subsequently made strenuous efforts to ensure Zweig's contributions were recognized.

and two down quarks, with a total charge of zero. Beta radioactivity could now be understood to involve the conversion of a down quark in a neutron into an up quark, turning the neutron into a proton, with the emission of a W⁻ particle.

Hints that there might be a fourth quark emerged in 1970. This was a heavy version of the up quark with a charge of $+\frac{2}{3}$, and was called 'charm'. It was now understood that the neutrino was paired with the electron (thus it is now called the electron neutrino). The muon neutrino was discovered in 1962. It seemed possible that the elementary building blocks of material substance consisted of two 'generations' of matter particles. The up and down quarks, the electron and electron neutrino formed the first generation. The charm and strange quarks, muon and muon neutrino formed a heavier second generation.

Most physicists were generally sceptical that a fourth quark was needed. But when another new particle, called the J/ψ, was discovered in 1974 simultaneously at Brookhaven National Laboratory in New York and the Stanford Linear Accelerator Center (SLAC) in California, it was realized that this must be a meson formed from charm and anti-charm quarks. Here was physical evidence that the charm quark exists. The scepticism vanished.

Such was the physicists' commitment to the emerging standard model of particle physics that when the discovery of yet another, even heavier, version of the electron – called the tau – was announced in 1977, it was quickly accommodated in a third generation of matter particles. American physicist Leon Lederman found the upsilon at Fermilab in August 1977, a meson consisting of what had by then come to be known as a bottom quark and its anti-quark. The bottom quark is a heavier, third-generation version of the down and strange quarks with a charge of $-\frac{1}{3}$.

The discoveries of the top quark and the tau neutrino were announced at Fermilab in March 1995 and July 2000 respectively. Together they complete the heavier third generation of matter particles, which consists of the top and bottom quarks, the tau and tau neutrino. Although further generations of particles are not impossible, there is some reasonably compelling experimental evidence to suggest that three generations is all there is.

Asymptotic freedom and the colour force

The quark model was a great idea, but at the time these particles were proposed there was simply no experimental evidence for their existence. Gell-Mann was himself rather cagey about the status of his invention. He had argued that the quarks were somehow 'confined' inside their larger hosts and, wishing to avoid getting bogged down in philosophical debates about the reality or otherwise of particles that could never be seen, he referred to them as 'mathematical'.

But experiments carried out at SLAC in 1968 provided strong hints that the proton is indeed a composite particle, made up of even smaller, more elementary constituents. It was not clear that these constituents were necessarily quarks, and the experimental results suggested that, far from being held together tightly, they were actually rattling around inside the proton as though they were entirely free. How could this be squared with the notion of quark confinement?

This puzzle was cleared up in 1973 by Princeton theorists David Gross and Frank Wilczek, and independently by David Politzer at Harvard. When we imagine a force acting between two particles, we tend to think of examples such as gravity or electromagnetism, in which the force grows stronger as the particles move closer together. But the strong nuclear force doesn't behave this way. The force exhibits what is known as *asymptotic freedom*. In the asymptotic limit of zero separation between two quarks, the particles feel no force and are completely 'free'. As the separation between them increases beyond the boundary of the proton or neutron, however, the strong force tightens its grip.

It is as if the quarks are fastened to each end of a piece of strong elastic. When they are close together, the elastic is loose. There is little or no force between them. But as we try to pull the quarks apart, we begin to stretch the elastic. The force increases the harder we pull.

Building on earlier work, Gell-Mann, German theorist Harald Fritzsch and Swiss theorist Heinrich Leutwyler now developed a quantum field theory of the strong nuclear force. In addition to the quark 'flavours' – up, down, strange, etc. – they introduced a new variable which they called 'colour'. Each quark can possess one of three different colour 'charges' – red, green or blue.

Baryons are formed from three quarks each of a different colour, such that their total 'colour charge' is zero and the resulting particle is 'white'. For example, a proton may consist of a blue up quark, a red up quark and a green down quark. A neutron may consist of a blue up quark, a red down quark and a green down quark. The mesons, such as pions and kaons, consist of coloured quarks and their anti-coloured anti-quarks, such that the total colour charge is zero and the particles are also 'white'.

In this model, quarks are bound together by a 'colour force', carried by eight massless particles called gluons, which also carry colour charge and, like the quarks, are confined inside the hadrons. Gell-Mann called the theory *quantum chromodynamics*, or QCD.

In essence, this completes the standard model of particle physics. The model consists of three generations of matter particles, a collection of force particles and the Higgs boson (Figure 2). The interactions of these particles are described by a combination of electro-weak field theory and QCD. The electro-weak theory is itself a combination of weak force theory (sometimes referred to as quantum flavour dynamics, or QFD)* and QED, their distinction forced as a result of the Higgs mechanism. We can therefore think of the standard model as the combination QCD × QFD × QED.

There is, at the time of writing, no observation or experimental result in particle physics that cannot be accommodated within this framework.** It is not the end, however, as we will see.

The construction of mass

I have a heavy glass paperweight on the desk in front of me. Where, exactly, does the *mass* of this paperweight reside?

The paperweight is made of glass. It has a complex molecular structure consisting primarily of a network of silicon and oxygen atoms bonded together. Obviously, we can trace its mass to the protons and neutrons which account for 99 per cent of the mass of every silicon and

---

* This might suggest that the theory applies only to quarks, but the different kinds of leptons (electron, muon, tau and their corresponding neutrinos) are also sometimes referred to as 'flavours'.

** Provided, that is, that the particle recently discovered at CERN proves to be the standard model Higgs boson.

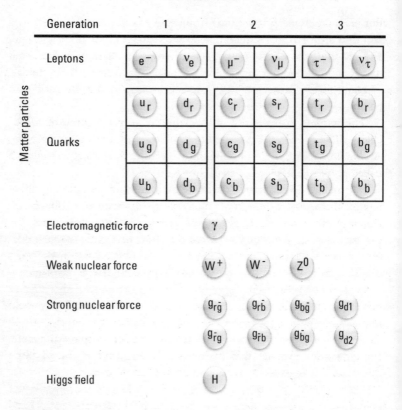

Figure 2 The standard model of particle physics describes the interactions of three generations of matter particles through three kinds of force, mediated by a collection of field particles or 'force carriers'. Note that only 'left-handed' leptons and quarks experience the weak nuclear force. The masses of the matter and force particles are determined by their interactions with the Higgs field.

oxygen atom in the structure. Not so very long ago, we might have stopped here, satisfied with our answer.

But we now know that protons and neutrons are made of quarks. So, we conclude that the mass of the paperweight resides in the cumulative masses of the quarks from which they are composed. Right?

Wrong. Because the quarks are confined, it's quite difficult to determine their masses, but it is known that they are substantially smaller and lighter than the protons and neutrons that they comprise.

For example, the Particle Data Group quotes mass ranges for both up and down quarks. If we pick masses at the higher end of these ranges and add the masses of two up quarks and a down quark, we get a result that represents about 1 per cent of the mass of a proton.

Hang on a minute. If 99 per cent of the mass of a proton is not to be found in its constituent quarks, where is it?

The answer to this question demands an understanding of colour charge. The Ancient Greeks used to claim that nature abhors a vacuum. A contemporary version of this aphorism might read: 'nature abhors a naked colour charge'.

In principle, the energy of a single isolated quark is infinite, which is why individual quarks have never been observed. It is much less expensive in energy terms to mask the colour charge that would be exposed by an individual quark either by pairing it with an anti-quark of the corresponding anti-colour or combining it with two other quarks of different colour such that the net colour charge is zero.

However, even within the confines of a proton or neutron, it is not possible to mask the exposed colour charges completely. This would require that nature somehow pile the quarks directly on top of one another, so that they all occupy the same location in space and time. But quarks are quantum wave particles, and they can't be pinned down this way – Heisenberg's uncertainty principle forbids it.

Nature settles for a compromise. Inside a proton or neutron the colour charges are exposed and the energy – manifested in the associated gluons that are exchanged between them – increases. The increase is manageable, but it is also substantial. The energy of the gluons inside the proton or neutron builds up, and although the gluons are massless, through $m = E/c^2$ their energy accounts for the other 99 per cent of the particle's mass.

And there you have it. For centuries we believed that it would one day be possible to identify the ultimate constituents of material substance, the ultimate 'atoms' of matter from which everything is constructed. This was Dirac's dream. We assumed that such constituents would possess certain characteristic physical properties, such as the primary quality of mass.

What the standard model of particle physics tells us is far stranger and, consequently, far more interesting. There do appear to be ultimate constituents (at least for now), and they do have characteristic physical properties, but mass is not really one of them. Instead of mass we have interactions between constituents that would otherwise be massless and the Higgs field. These interactions slow the particles down, giving rise to behaviour that we interpret as mass. As the constituents combine, the energy of the massless force particles passing between them builds, adding to the impression of solidity and substance.

Nobody said that science would deliver a description of empirical reality that was guaranteed to be easily comprehensible. But it is nevertheless rather disconcerting to have the rug of our common experience of light and matter pulled from beneath our feet in this way.

4

Beautiful Beyond Comparison

Space, Time and the Special and General Theories of Relativity

The theory is beautiful beyond comparison. However, only one colleague has really been able to understand it …

Albert Einstein[1]

When Newton published his classic work *Philosophiæ Naturalis Principia Mathematica* (*The Mathematical Principles of Natural Philosophy*) in 1687, he defined an authorized version of reality that was to prevail for more than two hundred years. Newton's mechanics became the basis on which we sought to understand almost everything in the physical world. There appeared to be no limits to its scope, from the familiar objects of everyday experience here on earth to objects in the furthest reaches of the visible universe.

But Newton had been obliged to sweep at least two fundamental problems under the carpet. The first of these appeared to be largely philosophical, and therefore (it could be argued) a matter of personal preference. The second seemed no less philosophical but was more visibly physical, and disconcerting. It would fall to Einstein, and those physicists and philosophers who inspired him and on whose work he built, to resolve these problems from within his special and general theories of relativity.

Einstein's theories of relativity were radically to transform how we seek to comprehend space and time, and the ways in which space and time respond to the presence of material substance.

After Einstein, reality would never be quite the same again.

Newton's bucket

The first problem Newton had to confront concerned the very nature of space and time. Are these things aspects of an independent physical reality? Do they exist independently of objects and of perception or measurement? In other words, are they 'absolute' things-in-themselves?

Take your eyes away from this book and look around you. What do you see? That's easy. You see objects in your immediate environment – perhaps chairs, a table, a TV in the corner of the room. These, you conclude, are objects in space.

But what, precisely, *is* space? Can you see it? Can you touch it? Well, no, you can't. Space is not something that we perceive directly. We perceive objects, and these objects have certain relations with one another which we might be tempted to call spatial relations, but space itself does not form part of the content of our direct experience. Our interpretation of the objects as existing in a three-dimensional space is the result of a synthesis of sense impressions in our brains translated by our minds.*

Similarly, you can't reach out and touch time. Time is not a tangible object. Your sense of time would seem to be derived from your sense of yourself and the objects around you changing their relative positions, or changing their nature, from one type of thing into another.

The pragmatists among us shrug their shoulders (again). So what? Just because we can't perceive these things directly doesn't mean they aren't 'real'. Newton was inclined to agree, although he was willing to acknowledge the essential relativity of space and time in our 'vulgar' experience. Objects move towards or away from each other, changing their relative positions in space and in time. This is relative motion, occurring in a space and time defined only by their relationships to the objects of reality.

But Newton's mechanics demanded an *absolute* motion. He argued that, although we can't directly perceive them, absolute space and time really do exist, forming a 'container' within which matter and energy can interact. Take all the matter and energy out of the universe, he

---

* Readers interested in exploring the nature of this synthesis should consult Michael Morgan's *The Space Between Our Ears: How the Brain Represents Visual Space*, published by Weidenfeld & Nicolson in 2003.

decreed, and the empty container would remain: there would still be 'something'.

Some of Newton's contemporaries (most notably his arch-rival, German philosopher and mathematician Gottfried Liebniz) were not satisfied. They preferred a more empiricist perspective, dismissing theoretical objects or structures that cannot be directly perceived as intellectual fantasies, akin to God or angels.

Newton believed that although absolute space and time cannot be directly perceived, we can nevertheless perceive phenomena that can only be explained in terms of an absolute space and time. To answer his critics, he devised a thought experiment to demonstrate this possibility. This is Newton's famous bucket argument.

Suppose we go into the garden.* We tie one end of a rope to the handle of a bucket and the other around the branch of a tree, so that the bucket is suspended in mid-air. We fill the bucket three quarters full with water. Now we turn the bucket so that the rope twists tighter and tighter. When the rope is twisted as tight as we can make it, we let go of the bucket and watch what happens.

The bucket begins to spin around as the rope untwists. At first, we see that the water in the bucket remains still, its surface flat and calm. Then, as the bucket picks up speed, the water itself starts to spin and its surface becomes concave – the water is pushed by the rotation out towards the circumference and up the inside of the bucket. Eventually, the rate of spin of the water catches up with the rate of spin of the bucket, and both spin around together.

Watching over your shoulder, Newton smiles. He wrote:

> This ascent of the water [up the side of the bucket] shows its endeavour to recede from the axis of its motion; and the true and absolute circular motion of the water, which is here directly contrary to the relative, discovers itself, and may be measured by this endeavour.[2]

Here's the essence of Newton's argument. The surface of the water becomes concave because the water is moving. This motion must be

---

* It'll have to be your garden. I live in a fourth-floor apartment.

either absolute or relative. But the water remains concave as its rate of spin relative to the bucket changes, and it remains concave when the water and the bucket are spinning around at the same rate. The concave surface cannot therefore be caused by the motion of the water relative to the bucket. It must be caused by the absolute motion of the water. Absolute motion must therefore exist. Absolute motion can only exist in absolute space. So, absolute space exists.

Have you spotted the potential flaw in Newton's argument yet?

If flaw it is, it is the assumption implicit in Newton's logic that if the concave surface of the water is not caused by motion relative to the bucket, then it surely cannot be caused by motion relative to the tree, the garden, me, you, the earth, the sun and all the stars in the universe.

Why not? Well, if this really were an example of motion relative to the rest of the universe, then by definition we cannot tell which component of such a system is stationary and which is moving. This is what relative motion means. Relative motion demands that if the bucket and the water in it were perfectly still, and we could somehow spin the entire universe around it, we would expect that the surface of the water would become concave.

Ridiculous! How could the entire universe have this kind of influence without appearing to exert a force on the water of any kind? Newton's assumption is surely valid, and absolute space must therefore exist.

Simultaneity and the speed of light

We might be inclined to think that the question of whether or not absolute space and time exist is a largely philosophical question. Indeed, absolute space and time suggest some kind of privileged frame of reference, a 'God's eye view', and Newton was quite willing to assign responsibility for it to God. We might conclude that the question of the existence or otherwise of such a privileged frame is interesting, likely to provoke interesting arguments, but ultimately unanswerable.

Einstein didn't think so. As he pondered this question whilst working as a 'technical expert, third class' at the Swiss Patent Office in Bern in 1905, he realized that it did, indeed, have an answer. He concluded that absolute space and time cannot exist, because the speed of light is a constant, independent of the speed of its source.

Although he was later to become a diehard realist, the youthful Einstein was greatly influenced by empiricist philosophy; particularly that of the Austrian physicist Ernst Mach. Mach had criticized Newton's concepts of absolute space and time and dismissed them as useless metaphysics.

Einstein was also something of a rebel, ready and willing to challenge prevailing opinions. He was at this time completely unknown to the academic establishment, and could therefore publish his outrageous ideas without fear of putting his (non-existent) academic reputation at risk.

In shaping his special theory of relativity, Einstein established two fundamental principles. The first, which became known as the principle of relativity, asserts that observers who find themselves in states of relative motion at different (but constant) speeds *must* observe precisely the same fundamental laws of physics.

This seems perfectly reasonable. For example, if an observer on earth makes a measurement to test Maxwell's equations and compares the result with that of another observer making the same measurement on board a distant spaceship moving away from the earth at high speed, then the conclusions from both sets of observations must be the same. There cannot be one set of Maxwell's equations for one observer and another set for space travellers.

We can turn this on its head. If the laws of physics are the same for all observers, then there is no measurement we can make which will tell us which observer is moving relative to the other. To all intents and purposes, the observer in the spaceship may actually be stationary, and it is the observer on earth who is moving away at high speed. We cannot tell the difference through physics.

The second of Einstein's principles concerns the speed of light. At the time he was working on special relativity, all attempts to find experimental or observational evidence for the ether – the universal medium that was supposed to fill the universe and support light waves – had failed. The most widely known of these attempts was conducted by American physicists Albert Michelson and Edward Morley. In 1887, they had made use of light interference effects in a device called an interferometer to search for small differences in the speed of light due to changes in the orientation of the earth relative to the ether, which was assumed to be stationary (and therefore absolute).

As the earth rotates on its axis and moves around the sun, a stationary ether would be expected to flow over and around it, creating an 'ether wind' that would change direction with the time of day and the season. Light travelling in the direction of such an ether wind could be expected to be carried along a little faster, just as sound waves travel faster in a high wind. Light travelling against the ether wind could be expected to be carried a little more slowly. As Michelson and Morley rotated their interferometer through $360°$, the changing orientation in relation to the ether wind would then be expected to manifest itself as a change in the measured speed of light. Any such differences in speed would be detected through subtle shifts in the observed interference fringes.

No differences could be detected. Within the accuracy of the measurements, the speed of light was found to be constant.

It seemed that the ether didn't exist after all. What's more, physicists had no choice but to conclude that the speed of light appears to be completely independent of the speed of the light source. If we measure the speed of the light emitted by a flashlight held stationary on earth,[3] then measure it again using the same flashlight on board a spaceship moving at high speed, we expect to get precisely the same answer.

Einstein concluded that he didn't need an ether. In his 1905 paper on special relativity he wrote: 'The introduction of a "light ether" will prove to be superfluous, inasmuch as the view to be developed here will not require a "space at absolute rest" endowed with special properties ...'[4]

In a bold move touched by genius, he now elevated the constancy of the speed of light to the status of a fundamental principle. Instead of trying to figure out *why* the speed of light is independent of the speed of its source, he simply accepted this as an established fact. He assumed the speed of light to be a universal constant and proceeded to work out the consequences.

One immediate consequence is that there can be no such thing as absolute time.

Suppose you observe a remarkable occurrence. During a heavy thunderstorm, you see two bolts of lightning strike the ground simultaneously, at points you label A (a certain distance to your left) and B (to your right).

A passenger on a high-speed train also observes the lightning. From your perspective, the train is moving from left to right, passing first point A and then point B. The lightning bolts strike as the passenger is passing through the midpoint of the distance between A and B.

The train is travelling *very* fast, at 100,000 miles per second, or about 54 per cent of the speed of light.* What does the passenger see?

Let's further suppose that the vantage point of the passenger at the midpoint between A and B is half a mile from both points. It will take about three millionths of a second for the light from the lightning flash at point B to cover this distance. But in that time the train has moved about a third of a mile to the right, *towards* point B. The light from the lightning bolt at point A now has a little further to travel before it reaches the passenger, and because the speed of light is constant, this takes a little longer. Consequently, the passenger sees the lightning strike at point B before she sees it strike at point A.

Because the speed of light is fixed, you and the passenger have different experiences. You perceive the lightning strikes to be simultaneous. The moving passenger sees the lightning strike point B first, then point A. Are the strikes actually simultaneous or not? Who is right?

You are both right. The principle of relativity demands that the laws of physics must be the same irrespective of the relative motion of the observer, and you cannot use physics to tell whether it is you or the passenger who is in motion.

We are left to conclude that there is no such thing as absolute simultaneity – events perceived to be simultaneous by you are not perceived to be simultaneous by the passenger. And as our very understanding of time is itself based on the notion of simultaneity between events, there can be no 'real' or absolute time. Something's got to give. You and the passenger perceive events differently because time is relative.

The dilation of time

Okay, so perhaps there's no such thing as absolute simultaneity and therefore no such thing as absolute time. But so what? All this business

---

* The speed of light in a vacuum is about 299.792 million metres per second, or 186,282 miles per second.

with lightning flashes and speeding trains just means that different observers moving at different relative speeds *perceive* things differently. It doesn't mean that time is *actually* different. Does it?

Let's imagine that the passenger has set up a little experiment to measure the speed of light on board the train. A small light–emitting diode (LED) is fixed to the floor of the carriage. This flashes once, and the light is reflected from a small mirror on the ceiling, directly overhead. The reflected light is detected by a small photodiode placed on the floor alongside the LED. Both LED and photodiode are connected to an electronic box of tricks that allows the passenger to measure the time interval between the flash and its detection. The height of the carriage is accurately and precisely known. From the distance travelled and the time it takes, the passenger can calculate the speed of the light.

For the sake of argument, let's assume that the height of the carriage is 12 feet, and the passenger carries out her first experiment when the train is stationary. The light makes a round trip of 24 feet (0.0045 miles) from floor to ceiling and back again. The passenger determines that this takes about 24.1 billionths of a second. The round-trip distance divided by the time taken gives a speed of light of 186,722 miles per second. Not a bad estimate.

The train then accelerates to a constant speed of 100,000 miles per second. Now, from your perspective as a stationary observer watching from the platform, the light from the LED doesn't travel vertically up and down. In the time it takes for the light to travel upwards towards the ceiling, the train has moved forward. It continues to move forward as the light travels back down to the floor. From your perspective, the light path looks like a 'Λ', a Greek capital lambda.

You therefore judge that the light now has further to travel, and the round trip takes longer. This could be compensated if the light were to move a little faster, covering the longer distance in the same time. But Einstein says that the speed of light is a constant, independent of the speed of its source. So, we use Pythagoras' theorem to work out that it now takes 28.6 billionths of a second for the light to make its round trip from floor to ceiling and back.[5]

As far as the passenger is concerned, nothing has changed. How is this possible? There is only one conclusion. From your perspective as a stationary observer, *time slows down on the moving train*. This is not like some optical illusion created by looking at identical figures against a changing perspective. The difference is not apparent, it is real. Muons in cosmic rays move at 99 per cent of light speed and decay five times more slowly than slow-moving muons created in the laboratory. When we put an atomic clock on board an aircraft and fly it from London to Washington DC and back, it loses 16 billionths of a second due to relativistic time dilation compared to an identical stationary atomic clock left behind at the UK's National Physical Laboratory.[6]

Those looking for a more practical application of special relativity need look no further than the Global Positioning System (GPS), used to navigate from place to place on land, sea and through the air. The GPS receiver in your smartphone or car navigation system uses signals from a network of orbiting satellites to triangulate and pinpoint your position to within a range of five to ten metres in a matter of seconds. Atomic clocks on board each satellite keep time to within 20–30 billionths of a second.

The satellites travel around their orbits at speeds of the order of 14,000 kilometres per hour. Relativistic time dilation causes the atomic clocks to run slow when compared with stationary clocks on earth: they lose about seven thousandths of a second each day. If this effect were not anticipated and corrected, the error in position would accumulate at the rate of a couple of *kilometres* per day! A bit of a problem if you're trying to find the location of your next meeting or a good restaurant.[7]

The contraction of length

What about space?

Let's now measure the length of the train. That's easy. Whilst it is stationary we calculate its length using the standard metre, made of platinum and iridium, which we borrow from the International Bureau of Weights and Measures in Sèvres, France. We find that the train is half a mile long.

But what if we now want to measure the length of the train as it moves along the track at a speed of 100,000 miles per second? We could try running alongside it with the standard metre, but this seems a bit impractical. Instead we ask our passenger to set up a different experiment. She now places the LED and photodiode detector at the back of the train and the mirror at the front. The LED flashes, light travels the length of the train and is reflected back by the mirror. She measures the time it takes to make one complete round trip, multiplies this by the speed of light and thus calculates the length of the train.

Our passenger first checks that everything is in order by doing the experiment on the train as it sits, stationary, by the platform. While you're busy running along the outside with the standard metre, she's discovering that the total round-trip time for the light flash is about 5.4 millionths of a second. This is consistent with a total round-trip distance of one mile, which means that the train is half a mile long. Perfect.

Now she performs the experiment on the moving train. She finds nothing untoward. Once again, however, from your perspective as a stationary observer, you see something rather different. As the light travels from the LED to the front of the train, the train is moving forward. The light therefore has to travel a distance that is a little longer than the actual length of the train. This extra distance is the distance covered by the train in the time it takes for the light from the LED to reach the mirror, given by the speed of the train multiplied by the time taken.

As the light reflected from the mirror travels back down the train, the back of the train is also moving forward, so the light travels a distance that is a little shorter than the actual length of the train. This foreshortening is given by the distance covered by the train in the time it takes for the light from the mirror to reach the photodiode. This is again calculated by the speed of the train multiplied by the time taken.

Okay, so we relax. The extra distance that the light has to travel on its way to the mirror is compensated by the shorter distance it travels on its way back. But, I'm sure you won't be surprised to learn, it doesn't quite work out that way. When we do the maths, we discover that the length of the moving train as judged by a stationary observer is now *shorter* than the length of the stationary train. The length of the moving train contracts by about 16 per cent, to something like 0.42 miles.[8]

This contraction in the lengths of objects in the direction of their motion is often called the Lorentz–FitzGerald contraction, named for Dutch physicist Hendrik Lorentz and Irish physicist George FitzGerald. They had both argued that the negative results of the Michelson–Morley experiment could be explained if the interferometer were physically contracting along its length in response to pressure from the ether wind, by an amount that precisely compensates the change in the speed of light, such that the speed of light is always measured to be constant.

Einstein's special theory of relativity makes the contraction no less physical. But instead of the object's atoms or molecules getting pushed closer together by the ether wind acting on the object in absolute space, it is instead *relative* space – the relative distance between objects as judged by a stationary observer – that is contracting.

There is, at present, no experimental evidence for length contraction of the kind predicted by special relativity. For 'everyday' objects moving at physically attainable speeds, the effects are too small to measure. Subatomic objects moving close to light speed would be expected to show these effects much more clearly, but the 'lengths' of these objects are not accessible to experiment (and, indeed, the dual wave particle nature of subatomic particles suggests that 'length' might not be a very meaningful property for such objects).

$E = mc^2$ and all that

Einstein's 1905 paper on special relativity was breathtaking in its simplicity yet profound in its implications. But he wasn't quite done. He continued to think about the consequences of the theory, and just a few months later he published a short addendum.

In this second paper, he considered the situation in which a body emits two bursts of electromagnetic radiation in opposite directions. Although Einstein had already published a paper speculating on the possibility of light quanta, he refrained from introducing this concept here; in any case he did not need it – what's important is that the radiation carries *energy* in some form away from the body.

Assuming the emitted bursts of radiation behave as waves, he examined the relationship between the energy of the system (object plus emitted radiation) from the perspective of the system's 'rest frame' and from the perspective of an observer moving at constant speed relative to the system. He found that the energies of the two situations are measurably different, and traced this difference to the kinetic energies of the body in the different inertial frames.*

---

* Kinetic energy is energy associated with the motions of objects.

After a little bit of algebraic manipulation, he concluded that if the total energy carried away by the radiation is E, then the mass of the body m must diminish by an amount E/c^2, where c is the speed of light. He continued: 'Here it is obviously inessential that the energy taken from the body turns into radiant energy, so we are led to the more general conclusion: The mass of a body is a measure of its energy content …'[9]

Today we would probably rush to rearrange the equation in Einstein's paper to give the iconic formula $E = mc^2$. But Einstein himself didn't do this. And he was not actually the first to note this possibility: five years earlier, French theorist Henri Poincaré had concluded that a pulse of light of energy E should have a mass given by E/c^2.[10]

Einstein's genius was once again to suggest that this is a general result – the inertia of a body (its resistance to acceleration), and hence the inertial mass of a body, is a measure of the amount of energy it contains. Although he was uncertain that this was something that could ever be subject to experimental test, he was prepared to speculate that the conversion of mass to energy might be observed in radioactive substances, such as radium.

Whilst $E = mc^2$ has become an iconic formula that demonstrates the equivalence of mass and energy, it is perhaps important to note that, as written, it is not universally applicable. It can't be applied to photons, for example, as photons are massless. If we put $m = 0$ into Einstein's equation we would erroneously conclude that photons possess no energy, either. This is because $E = mc^2$ is an approximation of a more general equation. From this general equation it is possible to deduce that photons possess energy given by their momentum multiplied by the speed of light.[11]

It is also tempting to push for one further conclusion regarding the nature of mass in special relativity. The equations can be manipulated to give an expression for the 'relativistic mass' of an object, which becomes larger and larger as the object is accelerated to speeds closer and closer to that of light.

The object is not literally increasing in size. It is mass as a measure of the object's resistance to acceleration that mushrooms to infinity for objects travelling at or near light speed.[12] This is obviously impossible, and often interpreted as the reason why the speed of light represents an ultimate speed which cannot be exceeded. To accelerate any object carrying mass to light speed would require an infinite amount of energy.

However, Einstein himself seems to have been cool on this interpretation, and the notion of 'relativistic mass' remains very dubious. In a 1948 letter to Lincoln Barnett, an editor at *Life* magazine who was working on a book about Einstein's relativistic universe, Einstein wrote:

> It is not good to introduce the concept of the [relativistic] mass M … of a moving body for which no clear definition can be given. It is better to introduce no other mass concept than the 'rest mass', m. Instead of introducing M it is better to mention the expression for the momentum and energy of a body in motion.[13]

Spacetime: travels in four dimensions

Depending on your perspective, time dilates and lengths contract. These bizarre effects of special relativity force us to abandon our cosy Newtonian conceptions of absolute space and time and appear to threaten us with chaos. If space and time are relative and depend on the speed of a moving observer, then how can we ever hope to make sense of the universe?

We need to stay calm. Einstein was led to the principle of relativity because we demand that the laws of physics should appear to be the same for all observers, irrespective of their state of relative motion. In other words, the laws of physics are *invariant*.

But how can this be if space and time are not invariant in the same way? The answer is relatively simple,* and was identified by Hermann Minkowski, Einstein's former maths teacher at the Zurich Polytechnic. On 21 September 1908, he opened his address to the 80th Assembly of German Natural Scientists and Physicians with these words:

> The views of space and time which I wish to lay before you have sprung from the soil of experimental physics, and therein lies their

---

* No pun intended.

strength. They are radical. Henceforth space by itself, and time by itself, are doomed to fade away into mere shadows, and only a kind of union of the two will preserve an independent reality.[14]

The fact that relative motion causes time to dilate and lengths to contract means that time intervals and spatial intervals (distances) are different for different observers. But, Minkowski realized, if space and time were to be combined in a four-dimensional *spacetime*, then intervals measured in this spacetime would remain invariant. In Minkowski spacetime, time multiplied by the speed of light – a product which has the units of distance – takes its rightful place alongside the three spatial dimensions of common experience.

But hang on a minute. If spacetime is invariant, does this mean it is also absolute? Newton was obviously wrong about absolute space and absolute time, but was he nearly right? Does an absolute spacetime restore the idea of an independently existing 'container' for mass and energy? Is it the existence of absolute spacetime that explains the absolute motion in Newton's bucket?

Many contemporary physicists and philosophers believe so. In *The Fabric of the Cosmos*, American theorist Brian Greene writes:

> Special relativity does claim that some things are relative: velocities are relative; distances across space are relative; durations of elapsed time are relative. But the theory actually introduces a grand, new, sweepingly absolute concept: *absolute spacetime*. Absolute spacetime is as absolute for special relativity as absolute space and absolute time were for Newton ...[15]

Gravity and the curvature of spacetime

We come, at last, to the second problem that Newton was obliged to sweep under the carpet: the problem of the origin of his force of gravity, which was required to act instantaneously between objects and at a distance.

It is such a common, everyday experience that it is hard to believe that Newton couldn't explain it. When I drop things, they fall. They do this because they experience the force of gravity. We are intimately familiar with this force. We struggle against it every morning when we get out of bed. We fight against it every time we lift a heavy weight. When we stumble to the ground and graze a knee, it is gravity that *hurts*.

In the clockwork universe described by Newton's laws of motion, force is exerted or imparted by one object impinging on another. But there is nothing slamming hard into the moon as it swoons in earth's gravitational embrace. There is nothing out at sea pushing the afternoon tide up against the shore. When a cocktail glass slips from a guest's fingers, it seems there is nothing to grab it and force it to shatter on the wooden floor just a few feet below.

Newton was at a loss. In the *Principia* he famously wrote:

> Hitherto we have explained the phænomena of the heavens and of our sea, by the power of Gravity, but have not yet assigned the cause of this power... I have not been able to discover the cause of those properties of gravity from phænomena, and I frame no hypotheses.[16]

Part of the solution to this riddle would come to Einstein during an otherwise average day at the patent office in November 1907, by which time he had been promoted, to 'technical expert, second class'. He later recalled: 'I was sitting in a chair in my patent office at Bern. Suddenly a thought struck me: If a man falls freely, he would not feel his weight.'[17]

Suppose you climb into an elevator at the top of the Empire State Building in New York City. You press the button to descend to the ground floor. The elevator is, in fact, a disguised interstellar transport capsule built by an advanced alien civilization. Without knowing it, you are transported instantaneously into deep space, far from any planetary body or star. There is no gravity here. Now weightless, you begin to float helplessly above the floor of the elevator.

What goes through your mind? Your sensation of weightlessness suggests to you that the elevator hoist cables have been suddenly cut and you're free-falling to the ground.

The aliens observing your reactions do not want to alarm you unduly. In true *Star Trek* style, these are beings of pure energy. They reach out with their minds, grasp the elevator/capsule in an invisible force field and gently accelerate it upwards. Inside the elevator, you fall to the floor. Relief washes over you. You conclude that the safety brakes must have engaged, and you have ground to a halt. You know this because, as far as you can tell, you're once more experiencing the force of gravity.

Einstein called it the 'equivalence principle'. The local experiences of gravity and of acceleration are the same.

The special theory of relativity is 'special' in the sense that it deals only with inertial frames at rest or moving at constant velocity relative to one another. Einstein had sought ways to deal theoretically with situations involving acceleration, and the equivalence principle suggested a strong connection between acceleration and gravity. But it would take him another eight years to figure out precisely how the connection works.

The equivalence principle suggested at least one consequence, however. In an elevator that is rapidly accelerating upwards, a beam of light emitted from one side of the elevator would strike the other side slightly lower down, because in the time it takes for the light to travel from one side to the other, the elevator has moved upwards. The acceleration has caused the light path to bend. And if acceleration cannot be distinguished from gravity, then we would expect to see precisely the same result if the elevator were stationary in a gravitational field. Einstein concluded that light is bent near large gravitating objects, such as stars and planets.

On its own, this was not a particularly astounding revelation. If, like Newton, we were to assume that light is composed of tiny corpuscles, then we would expect them to carry a small mass and so be affected by a gravitational field. Newton's corpuscular theory actually predicts a shift in the light from distant stars passing close to the sun of about 0.85 arc seconds. As we now know, photons are massless but they still possess momentum and energy, and this makes them susceptible to gravitational forces (because $'m' = E/c^2$).

But there is a second effect, as Einstein discovered. It originates in the understanding that light travels in straight lines, but what constitutes a straight line has to be amended in the vicinity of a large gravitating object.

Our general experience of distances and lines is formed by our local geometry, which is 'flat' or Euclidean. In Euclidean geometry the shortest distance between two points is obviously the straight line we can draw from one to the other. But what is the shortest distance between London and Sydney, Australia? The answer, of course, is 10,553 miles. But this distance is not, in fact, represented by a straight line. The surface of the earth is curved, and the distance between two points on such a surface is not a straight line, it is a curved path called a *geodesic*.

Now comes a leap of imagination that is truly breathtaking. What if the space near a large mass isn't 'flat'? What would happen if it is curved? Light following the 'straight line' path through a space that is curved would appear to bend. Einstein realized that he could finally eliminate the curious action-at-a-distance suggested by Newton's gravity by replacing it with curved spacetime. An object with mass (and hence energy) warps the spacetime around it, and objects straying close to it follow the 'straight line' path determined by this curved spacetime.

'Spacetime tells matter how to move; matter tells spacetime how to curve,' explained American physicist John Wheeler.[18] And this is a short, but eloquent, summary of Einstein's general theory of relativity.

Experimental tests of general relativity

Einstein estimated that the curvature of spacetime in the vicinity of the sun should make a further contribution to the bending of starlight, giving a total shift of 1.7 arc seconds.

This prediction was famously borne out by a team led by British astrophysicist Arthur Eddington in May 1919. The team carried out observations of the light from a number of stars that grazed the sun on its way to earth. Obviously such starlight is usually obscured by the scattering of bright sunlight by the earth's atmosphere, and can therefore only be observed during a total solar eclipse. Eddington's team recorded simultaneous observations in the cities of Sobral in Brazil and in São Tomé and Príncipe on the west coast of Africa. The apparent positions of the stars were then compared with similar observations made in a clear night sky.

Eddington was rather selective with his data, but his conclusions have since been vindicated by further observations. The shift is correctly predicted by general relativity.

Newton's theory of universal gravitation provided a powerful physical and mathematical underpinning of Kepler's three laws of planetary motion. Newton predicted that planets should describe elliptical orbits around the sun, each with a fixed perihelion – the point of closest approach of the planet to the sun. However, these points are not observed to be fixed – over time they *precess*, or rotate around the sun.

Much of the observed precession is caused by the cumulative gravitational pull of all the other planets in the solar system, and this can be predicted by Newton's theory. Collectively, it accounts for a precession in the perihelion of the planet Mercury of about 5,557 arc seconds per century. However, the observed precession is rather more, about 5,600 arc seconds per century, a difference of 43 arc seconds or 0.8 per cent.

Newton's gravity could not account for this difference, and other explanations – such as the existence of another planet, closer to the sun than Mercury – were suggested. General relativity predicts a contribution due to the curvature of spacetime of precisely 43 arc seconds per century.[*]

There is one further effect predicted by general relativity. This is the redshift in the frequency of light caused by gravity.

Once again, imagine for a moment that light consists of tiny corpuscles. Such corpuscles emitted from an object should be affected by the object's gravity – if the object were large enough, they could even be slowed down completely and pulled back. This seems all very reasonable, but we know that light doesn't consist of corpuscles (or, at least, ones with conventional mass). Reaching for a wave theory of light makes it difficult to understand the precise nature of the relationship that might exist between light and gravity. In classical wave theory, light is characterized by its frequency or wavelength, and once emitted, these properties are fixed. It simply wasn't obvious how gravity could exert an effect on the frequency of light waves.

Einstein figured out the solution in 1911. Gravity does not exert a direct effect on the frequency of a light wave, but it does have an effect on the spacetime in which the frequency is observed or measured. If, from an observer's perspective, time itself changes, then the frequency

---

[*] The perihelia of other planets are also susceptible to precession caused by the curvature of spacetime, but the contributions are much less pronounced.

of the wave (the number of up-and-down cycles per unit time) will change if measured against some standard external clock.

Einstein concluded that time should be perceived to slow down close to a gravitating object. A standard clock on earth will run more slowly than a clock placed in orbit around the earth. There are now two relativistic effects to be considered in relation to time. The atomic clock on board the plane from London to Washington DC loses 16 billionths of a second relative to the clock at the UK's National Physical Laboratory due to time dilation associated with the speed of the aircraft. But the clock *gains* 53 billionths of a second due to the fact that gravity is weaker at a height of 10 kilometres above sea level. In this experiment, the net gain is therefore predicted to be about 40 billionths of a second. The measured gain was reported to be 39 ± 2 billionths of a second.

What does this mean for light? As a light wave (or a photon) is emitted and travels away from a gravitating object, time speeds up as the effects of gravity reduce. There are now fewer up-and-down cycles per unit time. Seen from the reference frame of the source and a standard clock, those fewer up-and-down cycles per unit time are perceived as a shift to lower frequencies (longer wavelengths).

The effect is called the *gravitational redshift*. The light emitted from a gravitating object is shifted towards the red end of the electromagnetic spectrum as the effects of gravity weaken.

The opposite effect is possible. Light travelling towards a gravitating object will be blueshifted as the effects of gravity grow stronger. American physicists Robert Pound and Glen Rebka were the first to provide a practical earthbound test, at Harvard University in 1959. Gamma-ray photons emitted in the decay of radioactive iron atoms at the top of the Jefferson Physical Laboratory tower were found to be blueshifted by the time they reached the bottom, 22.5 metres below. The extent of the blueshift was found to be that predicted by general relativity, to within an accuracy of about 10 per cent (later reduced to 1 per cent in experiments conducted five years later).

Newton's bucket revisited: Gravity Probe B

Einstein pulled our understanding of space and time inside out. He had dismissed the notion of absolute space and time in his special theory of relativity. But the notion of an absolute spacetime persisted, as did the question posed by Newton's bucket, which suggested that absolute motion is possible.

When Einstein delivered the last of a series of four lectures on general relativity at the Prussian Academy of Sciences on 25 November 1915, he believed he had finally settled the matter. In the paper he wrote summarizing the theory, he claimed that its general relativistic principle 'takes away from space and time the last remnant of physical objectivity'.[19] In other words, he declared the defeat of the absolute and the triumph of the relative.

So, if the motion of the water in Newton's bucket is not absolute, to what, then, is it relative? We have established that it cannot be relative to the bucket itself. So it must be relative to the rest of the universe. And this means that if the bucket and the water in it were perfectly still, and we could somehow spin the entire universe around it, we would expect that the surface of the water would become concave. How can this be?

The answer is simply stunning. We could expect that the stationary water would be affected by the universe spinning around it because all the mass-energy in the universe collectively drags spacetime around with it as it spins. This was an effect first deduced from general relativity by Austrian physicists Josef Lense and Hans Thirring, known variously as *frame-dragging* or the Lense–Thirring effect. Frame-dragging means that there is no measurement we can make that would tell us if it is the water that is rotating in a stationary universe or the universe that is rotating around a stationary bucket of water. The motion of the water is relative.

So, is frame-dragging a real phenomenon? The answer is yes. On 24 April 2004, an exquisitely delicate instrument called Gravity Probe B was launched into polar orbit, 642 kilometres above the earth's surface. The satellite housed four gyroscopes, each with a 38-millimetre-diameter spherical rotor of fused quartz coated with superconducting niobium, cooled to -271°C. SQUIDs were used to monitor continually the orientations of the gyroscopes as the satellite orbited the earth. To eliminate unwanted torque on the gyroscopes, the satellite was rotated once every 78 seconds and thrusters kept it pointing towards the star IM Pegasi in the constellation of Pegasus.

Two effects were being measured. The fact that the earth curves the spacetime in its vicinity, and this causes the rotation axis of the gyroscopes to tilt (or precess) by a predicted 6,606 milliarc seconds per year (about 1.8 thousandths of a degree per year) in the plane of the satellite's orbit (that is, in a north–south direction). This precession is called *geodetic drift*, a phenomenon first identified by Dutch physicist Willem de Sitter in 1916.

The second effect is frame-dragging. As the earth rotates on its axis, it drags spacetime around with it in the plane perpendicular to the plane of the satellite orbit (in a west–east direction). This gives rise to a second precession of the gyroscopes, predicted to be 39.2 milliarc seconds per year.

The idea of an experimental test of general relativity based on satellite-borne gyroscopes had first been conceived in 1959. Over forty years elapsed from initial conception to launch, at a cost of $750 million. Data collection began in August 2004 and concluded about a year later. The project suffered a major disappointment when it was discovered that the gyroscopes were experiencing a substantial and unexpected wobble. Small patches of electrostatic charge on the rotors interacted with electrostatic charge on the inside of their housing, caused unexpected torque. These effects could be accounted for using an elaborate mathematical model, but at the cost of increased uncertainty in the final experimental results.

Consequently, analysis of the data took a further five years. The results were announced at a press conference on 4 May 2011. The geodetic drift, measured as a north–south drift in the orbital plane of the satellite, was reported to be 6,602±18 milliarc seconds per year. The west–east drift caused by frame-dragging was reported to be 37.2±7.2 milliarc seconds per year. The high (19 per cent) uncertainty in this last result was caused by the need to model the unexpected wobble.

Despite the uncertainty, this is still a very powerful experimental vindication of general relativity.

Einstein had argued that spacetime is relative. It owes its existence to matter and energy. Take all the matter and energy out of the universe and there would be no empty container. There would be nothing at all.

Interestingly, the debate did not end in 1916. There are further arguments to suggest that Einstein may have misinterpreted his own equations of general relativity. Today, many contemporary physicists and philosophers argue that Einstein was mistaken: spacetime may exist absolutely. It nevertheless *behaves* relatively, as the phenomenon of frame-dragging demonstrates. The debate is likely to run and run.

5

The (Mostly) Missing Universe

The Universe According to the
Standard Model of Big Bang Cosmology

We admittedly had to introduce an extension to the field equations that is not justified by our actual knowledge of gravitation.

Albert Einstein[1]

One of the most remarkable aspects of contemporary theoretical physics is its relatively new-found capacity to address questions that might be considered the preserve of high priests. Human beings possess a deep-rooted desire to understand their place in the universe. We have an innate need to fathom the seemingly unfathomable. Typically, what we are unable to fathom using observation, experiment and simple logic, the high priests attempt to explain through invention and the spinning of elaborate religious mythologies.

Today, the two principal building blocks that underpin our contemporary understanding of the physical world – relativity and quantum theory – are combined to tell the truly fascinating story of the origin and evolution of our universe. It is a story that is certainly no less remarkable than the creation myths of religious doctrine, and all the more remarkable because it happens to be 'true', at least for now, in the sense of the Veracity Principle.

Most readers will be already familiar with aspects of this modern creation story.

We now know that, insofar as the word 'began' is deemed appropriate, the universe began some 13.7 billion years ago in a 'big bang', a primeval quantum fluctuation of some kind that led to the creation of space, time and energy. What we now recognize as the four fundamental forces of nature disentangled themselves from the first, primeval force, in a series of what we might think of as phase transitions, much as steam condenses to water which freezes to ice.

Gravity was the first to be spun off, followed by the strong nuclear force, whose splitting triggered a short burst of exponential expansion of spacetime called *inflation*. Quantum fluctuations from this beginning of all things became imprinted by inflation on the large-scale structure of the universe we see today: a telltale thumbprint left at a cosmic crime scene. A subsequent phase transition separated the weak nuclear force from electromagnetism.

About 380,000 years after the big bang, primordial electrons latched themselves on to primordial atomic nuclei in a process called 'recombination'. The first neutral hydrogen and helium atoms were formed, releasing a flood of hot electromagnetic radiation to fill all of space.

The universe continued to evolve and expand, a fact belied by the simple observation that the night sky is largely dark.[2] The hot radiation released during recombination cooled as the universe expanded, and appears today in the form of microwaves with an average temperature of around -270.5°C, or 2.7 kelvin, almost three degrees above absolute zero. It is a cold remnant, an 'afterglow' of a tumultuous time in the history of the universe.

This cosmic microwave background (CMB) radiation was first detected in 1964. A succession of satellite surveys has mapped the CMB in exquisite detail, and provides much of the observational evidence on which theories of the origin and evolution of the universe are constructed. The quantum fluctuations that rippled through spacetime as the universe ballooned in size were the seeds for the subsequent formation of gas, clouds, stars, galaxies and clusters of galaxies. So, the pattern of points of light that we see in a night sky is a reflection of those quantum ripples from the dawn of time.

Despite this astonishing progress, the universe remains an almost complete mystery. But it remains a mystery for all the *right* reasons. More evidence from observational astronomy led physicists to conclude that there must exist an extraordinary form of matter presently unknown to the standard model of particle physics. We know next to nothing about this form of matter. Whatever it is, it cannot be affected by the electromagnetic force, since then it would become visible to us (in the form of radiation). It cannot be affected by the strong nuclear force, otherwise we would be able to observe its effects on visible matter. We know it exerts gravitational effects, and may also be susceptible to the weak force. It is utterly mysterious, and truly deserving of the name 'dark matter'.

There's more. Observational astronomy also suggests that the expansion of the universe is (rather counterintuitively) *accelerating*. Present theoretical structures can accommodate an accelerating expansion by assuming that the universe is filled with an invisible energy field, which has inevitably become known as 'dark energy'.

Dark matter and dark energy are no mere quirks. These are not mildly curious phenomena at the edges of our understanding waiting, like undotted i's and uncrossed t's, for the tidy pen strokes of explanation. When placed in a theoretical structure called the standard model of big bang cosmology, the most recent data from observational astronomy indicate that the density of dark matter represents about 22 per cent of all the mass-energy in the universe. The density of dark energy accounts for a further 73 per cent.

Visible physical matter and radiation – everything we can see in the universe and everything we are – accounts for just 5 per cent of everything there must be.

The universe is mostly missing.

Einstein's biggest blunder

The development of the ΛCDM (lambda: cold dark matter) model of the universe, also known as the standard model of big bang cosmology, is a triumph of modern physics. It will come as no surprise to learn that this is a development whose origins can be traced back to Einstein.

General relativity is all about the large-scale motions of planets, stars and galaxies and the structure of the universe within a four-dimensional spacetime. It deals with the universe in all its vastness.

At first, the task of applying general relativity to develop a theory of the entire universe (in other words, a cosmology) seems impossibly difficult. It's hard enough to keep track of the subtle interplay between gravitational forces and planetary motions within our own solar system. How then could it ever be possible to apply the theory to all the stars and galaxies in the observable universe?

The simple answer is: by making a few auxiliary assumptions. Although we can clearly see that the patterns of stars and galaxies are quite different in different parts of the night sky, we can nevertheless draw some simplifying conclusions about the 'coarse-grained' structure of the universe.

For one thing, there are no large patches of night sky that are completely devoid of starlight. The universe looks more or less the same in all directions, in the sense that we see roughly the same numbers of stars and galaxies, with roughly the same brightness. Secondly, the stars and galaxies that we see are not vastly different from one another in composition. There are certainly differences in the sizes of stars, galaxies and clusters of galaxies, and this leads to differences in their physical behaviour, but they are all made of the same kind of 'stuff', mostly hydrogen and helium.

So, we can assume that the universe is roughly uniform in all directions and uniform in composition. We must also further assume something called the *cosmological principle*, which states that stargazers on earth occupy no special or privileged position in the universe.* What we see from our vantage point on earth (or earth orbit) accurately reflects the way the universe appears from any or all such vantage points. What we see is a 'fair sample' of the universe as a whole.

With these assumptions in place, in 1917 Einstein applied the general theory of relativity to the entire universe. But he immediately hit a major problem. He expected that the universe that should emerge from his calculations would be consistent with prevailing scientific prejudice – a universe that is stable, static and eternal. What he got instead was a universe that is unstable and dynamic.

Gravity is the weakest of nature's forces, but it is cumulative and inexorable and acts only in one 'direction' – it attracts but does not repel. Einstein realized that the mutual gravitational attraction between all the masses in the universe would inevitably result in a universe that

---

* This sounds a bit like the Copernican Principle described in Chapter 1, but the Copernican Principle goes a lot further. The cosmological principle is in essence a kind of statistical assumption – our perspective from earth-bound or satellite-borne telescopes is representative of the entire universe. The Copernican Principle subsumes the cosmological principle, but goes on to insist that this perspective is due to the fact that the universe was not designed specifically with human beings in mind.

collapses in on itself. This was a disastrous result, quite inconsistent not only with prevailing scientific opinion but also arguably with simple observation. Several centuries of astronomy had yielded no evidence that all the stars in the universe were rushing towards each other in a catastrophic collapse.

This was a problem that was neither new nor a particular feature of general relativity. When applied to the universe as a whole, Newton's gravity also predicts a collapsing universe. Newton had resolved the problem by suggesting that God acts to keep the stars apart: '… and lest the systems of the fixed stars should, by their gravity, fall on each other mutually, he hath placed those systems at immense distances one from another'.[3] Einstein felt he needed something a little more scientific than this.

His solution was to modify arbitrarily the equations of general relativity as applied to the universe by introducing a 'cosmological constant', usually given the symbol Λ (lambda). This is the 'extension to the field equations' referred to in the title quotation.

In essence, the cosmological constant imbues space itself with a kind of anti-gravitational force, a negative pressure which increases in strength over longer distances. By carefully selecting the value of this constant, Einstein found that he could counterbalance the gravitational attraction that tended to pull everything together with a space that tended to push everything apart. The result was equilibrium, a static universe.

It was a relatively neat solution. Introducing the cosmological constant didn't alter the way general relativity works over shorter distances, so the successful predictions of the perihelion of Mercury and the bending of starlight were preserved. But it was, nevertheless, a rather unsatisfactory 'fudge', one that was 'not justified by our actual knowledge of gravitation'. There was no evidence for the cosmological constant, other than the general observation that the universe *seems* to be stable, and static.

Einstein found it all rather ugly and would come to regret his decision, as he later revealed to Ukrainian-born theoretical physicist George Gamow: 'When I was discussing cosmological problems with Einstein he remarked that the introduction of the cosmological term was the biggest blunder he ever made in his life.'[4]

The expanding universe

Einstein had taken great pains to ensure that the solutions to the gravitational field equations of general relativity yielded a universe that conformed to physical experience. But, of course, equations are just equations – the fact that they can be applied to physical problems doesn't necessarily mean that the only solutions are physically realistic or sensible ones.

In 1922, Russian physicist and mathematician Alexander Friedmann offered three models based on solutions of Einstein's field equations. These were essentially descriptions of three different kinds of 'imaginary' universe.

In the first, the density of mass-energy is high (lots of stars in a given volume of space) and spacetime is *expanding*, although the rate of expansion is modest. Such a universe is said to be 'closed': it would expand for a while before slowing, grinding to a halt and then turning in on itself and collapsing. In the second, the density of mass-energy is low (fewer stars) and the effects of gravity are insufficient to overcome expansion. Such a universe is said to be 'open', and would expand for ever.

In the third model, the density of mass-energy and the rate of expansion are finely balanced, such that gravity can never quite overcome the expansion. Such a universe is said to be 'flat'. The rate of expansion slows but it never stops. And a slow rate of expansion would give the appearance of a static universe.

Each of these different universes is characterized by the value of a density parameter, given the symbol Ω (omega), the ratio of the density of mass-energy to the critical value required for a flat universe. A closed universe has Ω greater than 1, an open universe has Ω less than 1 and a flat universe has Ω equals 1.

Friedmann's model universes were very different to Einstein's. They were dynamic, not static. Einstein initially rejected Friedmann's solutions as wrong, and was quickly obliged to retract when he realized that, mathematically speaking, the solutions were perfectly acceptable:

> I am convinced that Mr Friedmann's results are both correct and clarifying. They show that in addition to the static solutions to the field equations there are time varying solutions with a spatially symmetric structure.[5]

But the retraction referred only to the mathematical rigour of Friedmann's analysis. Einstein was convinced that the idea of an expanding universe had nothing to do with reality.

Tragically, Friedmann died in 1925. His expanding-universe solutions were independently rediscovered two years later by Belgian theorist (and ordained priest) Georges Lemaître. But Lemaître published his results in French in a rather obscure Belgian journal, and they attracted little attention.

Hubble's law

By 1931, everything had changed. Einstein was forced to accept that an expanding universe was not only possible, but appeared to describe the universe we inhabit. He publicly acknowledged that Friedmann and Lemaître had been right, and he had been wrong.

What had convinced him were the results of a series of observations reported by American astronomer Edwin Hubble and his assistant Milton Humason. The results pointed to a relatively unambiguous conclusion: most of the galaxies we observe in the universe are moving away from us.

Hubble had already radically transformed our understanding of the universe in the early 1920s. He had shown that what had appeared to be wispy patches of interstellar gas and dust – called nebulae – were in fact vast, distant galaxies of stars much like our own Milky Way. This revolutionary revision in our understanding that the Milky Way is but one of a huge number of galaxies not only greatly increased the size of the known universe; it also begged questions concerning a growing mystery surrounding their speeds.

Starting in 1912, American astronomer Vesto Slipher at Lowell Observatory in Flagstaff, Arizona, had used the Doppler effect to investigate the speeds of what were then still judged to be nebulae. The technique works like this. When we receive a wave signal (light or sound) from a moving object, we find that as the object approaches, the waves become bunched and their pitch (frequency) is detected to be higher than the frequency that is actually emitted. As the object moves away, the waves become spread or stretched out, shifting the pitch to lower frequencies. The effect is familiar to anyone who has listened to the siren of an ambulance or police car as it speeds past.

If we know the frequency that is emitted by the source, and we measure the frequency that is detected, then we can use the difference to calculate the speed at which the source is moving, towards or away from us, the receiver.

Stars are composed mostly of hydrogen and helium atoms. Hydrogen atoms consist of a single proton orbited by a single electron. Depending on its energy, the electron wave particle may be present in any one of a number of 'orbitals' inside the atom, forming discrete shapes or clouds of probability relating to where the electron actually is. Each orbital has a characteristic, sharply defined energy. Consequently, when an excited electron releases energy in the form of a photon, the frequency of the photon lies in a narrow range, determined by the difference in the energies of the two orbitals involved.

The result is an atomic *spectrum*, a sequence of 'lines' with each line representing the narrow range of radiation frequencies absorbed or emitted by the various electron orbital states inside the hydrogen atom. These frequencies are fixed by the energies of the orbitals and the physics of the absorption or emission processes. They can be measured on earth with great precision.

But if the light emitted by a hydrogen atom in a star that sits in a distant galaxy is moving relative to our viewpoint on earth, then the spectral frequencies will be shifted by an amount that depends on the speed with which the galaxy is moving.

In his observations, Slipher actually used two light frequencies characteristic of calcium atoms. He discovered that light from the Andromeda nebula (soon to be relabelled the Andromeda galaxy) is blueshifted (higher frequencies), suggesting that the galaxy is moving at high speed *towards* the Milky Way.* However, as he gathered more data on other galaxies, he found that most are redshifted (lower frequencies), suggesting that they are all moving away.

Hubble and Humason now used the more powerful 100-inch telescope at Mount Wilson near Pasadena, California, to gather data on more galaxies. What they found was that the majority of galaxies are indeed receding from us. Hubble discovered, in what appears to be an

---

* This is a Doppler shift, caused by the speed of motion of the light source, not to be confused with a gravitational shift due to the curvature of spacetime.

almost absurdly simple relationship, that the speed at which each galaxy is receding is proportional to the galaxy's distance. This is *Hubble's law*.[6]

The fact that most of the galaxies are receding from us does not place us in an especially privileged position at the centre of the universe. In an expanding universe it is spacetime that is doing the expanding, with every point in spacetime moving further away from every other point. The standard analogy is to think of the three-dimensional universe in terms of the two dimensions of the skin of a balloon. If we cover the deflated balloon with evenly spaced dots, then as the balloon is inflated the dots all move away from each other. And the further away they are, the faster they appear to be moving.

Lemaître had actually predicted this kind of behaviour in 1927, and had even derived a version of Hubble's law. But in an English translation of his 1927 paper that was published in 1931, all references to his derivation were inexplicably (and very carefully) excised.

If the universe is expanding, then simple logic suggests that we can 'wind the clock back' and conclude that it must have had an origin at some point in time. In 1927, Lemaître had not speculated on such a moment of 'creation', but he now went on to envisage this to involve what he called a 'primeval atom': essentially all the material substance of the universe compressed into a single atomic structure. He then likened the creation of the expanding universe to the process of radioactive disintegration.

There were plenty of problems still. Hubble's original work suggested that the universe is younger than the estimated age of the earth. But these problems were eventually resolved in the 1950s, most notably by Allan Sandage, another Mount Wilson astronomer, who would eventually come to be seen as Hubble's successor. Today, the value of the Hubble constant – the constant of proportionality between galactic speed and distance – is determined to be about 70 kilometres per second per megaparsec,* and the age of the universe is 13.7 billion years.

---

* A megaparsec is equivalent to 3.26 million light-years (a light-year is the distance that light travels in a year), or 30.9 billion billion kilometres.

The big bang

Lemaître's concept of a 'primeval atom' was obviously speculative. Indeed, it might be thought that any suggestion concerning a moment of creation extrapolated backwards from the large-scale structure of today's universe would remain for ever beyond science and therefore intangible.

But George Gamow thought differently. He realized that when aspects of atomic and nuclear physics are applied to the problem, the large-scale structure of today's universe inherently limits the creation possibilities. It was possible to say an awful lot more about the early universe.

The disintegration of Lemaître's primeval atom could not explain the relative abundances of hydrogen, helium and the sprinkling of other atoms in the universe. An atom whose nucleus contained all the protons and neutrons of the universe would indeed be unstable and would decay rapidly, but most nuclear fission reactions involve the *splitting* of nuclei rather than their utter fragmentation into individual protons and neutrons. It seemed rather more logical to start with an early universe consisting of primordial protons, neutrons and electrons and apply the principles of nuclear physics to work out what would most likely happen next.

With support from postgraduate student Ralph Alpher, Gamow did precisely this. Estimating the conditions likely to have prevailed in the first few minutes after the moment of creation, Alpher and Gamow successfully predicted the relative abundance of hydrogen and helium in today's universe.[7]

Predicting the abundance of hydrogen and helium in the universe may not sound like much of an achievement, but at this time the amount of helium was not precisely known, and this was big news. Through the work of Hubble and Humason, the evidence for an expanding universe was becoming overwhelming. But, although an expanding universe would seem to imply an inevitable beginning, the simple fact that it is expanding cannot be taken as evidence that it expanded from some origin, or that this origin represents the creation of all material substance and radiation.

Indeed, maverick English astronomer Fred Hoyle had developed an alternative explanation. Together with Austrian astronomers Thomas

Gold and Hermann Bondi, Hoyle developed a 'steady-state' model of the universe in which the expansion of spacetime is eternal and new matter is constantly emerging from a hypothetical 'C-field', or creation-field, which pervades the universe.

Now this might seem rather far-fetched, but creation of new matter at a rate of one atom per year in a volume of space equal to St Paul's Cathedral in London is all that is required to maintain a universal steady state. Over time, the new matter would be drawn together by gravity, forming first gas clouds, then stars and galaxies.

Hoyle rejected the theory that there could ever have been a 'moment of creation'. In a radio programme broadcast by the BBC in 1949, he introduced a new term – 'big bang' – intended to disparage the idea:

> On scientific grounds this big bang hypothesis is much the less palatable of the two [i.e. less palatable compared to the steady-state model]. For it is an irrational process that cannot be described in scientific terms ... On philosophical grounds too I cannot see any good reasons for preferring the big bang idea. Indeed it seems to *me* ... a distinctly unsatisfactory notion, since it puts the *basic* assumption out of sight where it can never be challenged by a direct appeal to observation.[8]

The importance of Alpher and Gamow's work derives from the fact that this was the first attempt to describe the immediate aftermath of the big bang in scientific terms. The conclusion was that if the universe had begun in a big bang, then the operation of the principles of nuclear physics could explain the relative abundance of hydrogen and helium, which together constitute 99.99 per cent of all visible matter.

The cosmic microwave background radiation

Predicting the relative abundance of hydrogen and helium was an encouraging result, but not one that could be considered to constitute a proof of the big bang model. Hoyle, for one, was unimpressed.

Alpher had become rather frustrated that the model he had developed with Gamow didn't predict the synthesis of elements heavier than helium.* In the meantime, Gamow had forged ahead on other aspects of early post-big-bang physics. In the summer of 1948 he sent Alpher a manuscript of a paper he had recently submitted to the British journal *Nature*. The paper was concerned with the densities of matter and radiation up to the point of recombination.

At the temperatures and pressures prevailing inside the primordial fireball, all material substance would be present in the form of hot plasma. Protons (hydrogen nuclei), clusters of two protons and two neutrons (helium nuclei) and electrons would move freely within such a plasma, exchanging electromagnetic radiation (photons) between them. As the universe expanded and cooled, the temperature would eventually drop to around 3,000°C, at which point recombination would occur. Electrons would be captured by the hydrogen and helium nuclei to form neutral atoms. After recombination, the hot radiation that had been exchanged between the constituents of the plasma would be released. The previously opaque universe would become transparent.

Let there be light. Literally.

Alpher and fellow physicist Robert Herman realized that Gamow's estimates of the densities of matter and radiation were seriously wrong, and sent a telegram to Gamow at the US atomic weapons research laboratory at Los Alamos, where he was working through the summer months. Gamow judged that it was too late to retract his paper, and urged Alpher and Herman to submit a note correcting his error for publication in the same journal.

Alpher and Herman went a little further. They took the opportunity in this short note to explain that the radiation released after recombination would have persisted to the present day. They estimated that this cosmic background radiation would have an average temperature just five degrees above absolute zero (5 kelvin). Although they didn't say so, it was implicit in their proposal that the radiation would be present in the form of microwaves.

---

* Although some heavier elements were formed in the big bang, much of the synthesis responsible for the distribution of elements in the universe is now understood to take place in the interiors of stars and during cataclysmic stellar events, such as supernovae. Hoyle played a significant role in working out the mechanisms of stellar nucleosynthesis. There's more on this in Chapter 11.

Now this was an out-and-out prediction. Unlike the relative abundances of hydrogen and helium, the existence of cosmic microwave background radiation had not been anticipated. And this was a prediction that only a model based on a hot big bang could make.

The prediction was largely ignored. In his BBC radio broadcast Hoyle had demanded a 'direct appeal to observation', of a kind of which he believed the big bang model to be incapable. Yet here was a simple test. Did the CMB radiation exist, or not?

Alpher and Herman later explored possible reasons why their work did not attract much attention:

> There was the just mentioned age problem; by the 1950s new values of Hubble's parameter had eliminated this. Also, some scientists had a predilection toward a steady-state universe. Finally, there was the above-mentioned view of Gamow's work and, perhaps by association, our work [Gamow's reputation for playfulness meant that his work wasn't always taken seriously].[9]

It is also a fact that whilst Alpher and Herman had worked in an academic institution, cosmology hadn't yet established itself as an important scientific discipline, and neither physicist belonged to the recognized communities of astrophysicists and astronomers. By the time cosmology had become more fashionable, both Alpher and Herman had left academia to pursue careers in industrial science.

Consequently, when in the summer of 1964 Princeton physicist Robert Dicke suggested: 'Wouldn't it be fun if someone looked for this radiation?' he was unaware of Alpher and Herman's earlier prediction.* Dicke had independently rediscovered the possible existence of the CMB and assigned the task of working out how to detect the radiation to two young radio astronomers, Peter Roll and David Wilkinson. Turning to Jim Peebles, a theorist from Manitoba, he said: 'Why don't you go and think about the theoretical implications?'[10]

---

* And, indeed, of several other, similar, predictions that Gamow, Alpher and Herman had published in the intervening years.

Peebles went home and thought about it. He reinvented the big bang model that Gamow, Alpher and Herman had developed and used it to predict a CMB radiation with a temperature about ten degrees above absolute zero. But when he submitted this work for publication, it was rejected, on the basis that Alpher, Herman and Gamow had already covered this ground some years before.

When, in early 1965, Dicke, Peebles, Wilkinson and Roll assembled for a lunchtime meeting to discuss the design of an apparatus to detect the CMB radiation, Dicke was interrupted by a phone call from radio astronomer Arno Penzias at Bell Laboratories in Holmdel, New Jersey. Penzias had caught sight of a preliminary manuscript on the CMB radiation by Dicke and Peebles and realized that this was a solution to a problem that he and fellow Bell Labs radio astronomer Robert Wilson had been puzzling over for some months. Dicke put the phone down. 'Well, boys,' he said. 'We've been scooped!'[11]

Dicke, Roll and Wilkinson piled into a car and drove the thirty-odd miles to Holmdel. They found that Penzias and Wilson, accomplished radio astronomers, had constructed a highly sensitive radiowave detector with a twenty-foot horn antenna. However, when they had first switched it on, they had picked up a persistent and annoying hiss of microwave radiation that appeared to come uniformly from all directions in the sky. It was only when Penzias had seen Dicke and Peeble's manuscript at the beginning of 1965 that the penny dropped.

The radiation that Penzias and Wilson had stumbled on quite by accident fitted closely with the Princeton physicists' predictions, indicating a radiation temperature of about 3 kelvin. Penzias and Wilson and the Princeton group published companion papers announcing the discovery in May 1965.

The big bang was no longer idle theoretical speculation. It was observational fact.

The flatness and horizon problems

The discovery of the CMB radiation was a triumph for big bang cosmology, and the alternative steady-state model disappeared from discussion almost overnight. We now knew that the universe had 'begun' in a tiny cosmic fireball which had expanded to form the universe we can observe today. It was, admittedly, a complex creation story, involving aspects of quantum physics, nuclear physics, the dynamics of gas clouds, star formation, stellar nucleosynthesis, galaxy and galactic cluster formation, supernovae and the formation of heavier elements, more gas clouds, more stars (this time with planetary systems), earth, life and us. It was obvious that there were still some gaps to fill.

But the big bang model also had some problems. Einstein had based his initial calculations on the perfectly reasonable assumptions that the universe is more or less the same in all directions (it is 'isotropic'), and that the stars and galaxies are not vastly different from one another in composition (the universe is homogeneous). What's more, the universe that we observe today is 'flat', meaning that there are three spatial dimensions, in which Euclidean geometry applies – the angles of a triangle add up to 180°, the square of the hypotenuse of a right-angled triangle is equal to the sum of the squares of the other two sides (Pythagoras' theorem), the ratio of the area of a circle and the square of its radius is equal to π, and so on.

This might seem a statement of the blindingly obvious, but the 'flatness' of space depends critically on the density of mass-energy in the universe.

The problem was that, in order to create a perfectly flat universe, with a density parameter Ω of precisely 1, the conditions that prevailed during the big bang would have had to have been rather special, and extremely fine-tuned. Dicke had first raised his hand to declare that this was a problem in the 1960s, and in 1979 he and Peebles published an important paper on the subject. Physicists grow increasingly nervous when confronted with inexplicable coincidences, as these are often open to misinterpretation as evidence of design.[*]

Dicke presented his arguments in a lecture he delivered at Cornell University in 1978. Sitting in the audience was a young American postdoctoral researcher called Alan Guth, who was struck by the implications:

> The way [Dicke] phrased the problem was that the expansion rate of the universe at one second after the big bang had to be fine-tuned to

---

[*] Pope Pius XII had pronounced in 1951 that the big bang model was consistent with the official doctrine of the Roman Catholic church.

the accuracy of, I think the number he gave was one part in 10^{14} [one hundred thousand billion]. And if the rate were one part in 10^{14} larger than it actually was, the universe would fly apart without galaxies ever forming. And if the universe were expanding at a rate of one part in 10^{14} slower, the universe would have long ago collapsed.[12]

There was another problem. Measurements of the CMB radiation confirmed its astonishing uniformity in temperature across the sky. Now we know that if we place a hot object in contact with a cool object, the temperatures of both objects will equilibrate and become uniform. Such uniformity in temperature implies a dynamic exchange of energy, which can take the form of radiation, between objects that are said to be in 'causal contact'.

But the last moment that such an exchange can have taken place was at the moment of recombination, about 380,000 years after the big bang. After that moment, the radiation was freed from matter and would interact with it very rarely. The trouble was, the finite speed of light prevents the entire universe from being causally connected. There would be many regions of the universe more distant than can be reached by radiation travelling at the speed of light, and so these regions could not be in causal contact. There was really no good reason why the temperature of the CMB radiation should be so uniform. This was called the horizon problem.

The resolution of both the flatness and horizon problems would involve a marvellous fusion of theories of particle physics and cosmology.

Cosmic inflation

The discovery of the weak neutral currents at CERN and Fermilab in 1973 lent considerable credibility to the electro-weak theory that Weinberg and Salam had constructed using the Higgs mechanism. At the time, there was no particle accelerator available that could re-create the energies at which the electromagnetic and weak nuclear forces would 'melt' back into the single electro-weak force, carried by massless bosons, from which they had originated. But it became entirely reasonable to suppose that such energies would have prevailed in the first moments after the big bang.

This is now known as the 'electro-weak epoch'. As the universe cooled, the background Higgs field 'crystallized' and the higher symmetry of the electro-weak force was broken (or, more correctly, 'hidden'). The massless bosons of electromagnetism (photons) continued unimpeded, but the weak force bosons interacted with the Higgs field and gained mass to become the W and Z particles.

There are close physical analogies between the transitions between these early epochs and the phase transitions of common, everyday substances, such as water. As temperatures fall, steam condenses to liquid water, and the symmetry is reduced as previously free-moving water molecules become connected together in large-scale network structures characteristic of the liquid. As temperatures fall again, water freezes to ice, further reducing the symmetry as the water molecules become locked in a solid crystal structure.

It required no great leap of logic to suggest that, in an even earlier epoch, the electro-weak and strong nuclear forces would have been similarly joined in a single 'electro-nuclear' force.

In 1974, Weinberg, American theorist Howard Georgi and Australian-born physicist Helen Quinn had shown that the strengths of the interactions of all three particle forces become near equal at energies between a hundred billion and a hundred million billion giga electron-volts (GeV).* These energies, corresponding to temperatures of around ten billion billion billion (10^{28}) degrees, would have been prevalent at about a hundred million billion billion billionth (10^{-35}) of a second after the big bang.

In this 'grand unification epoch', the strong nuclear force and electro-weak force melt into a single electro-nuclear force. All force carriers are identical and there is no mass, no electrical charge, no quark flavour (up, down) or colour (red, green, blue). Breaking this even higher symmetry in theory requires more Higgs fields, crystallizing at higher temperatures and so forcing a divide between quarks, electrons and neutrinos and between the strong and electro-weak forces.

---

* More recent evaluation puts this energy somewhere in the region of 200,000 billion GeV.

One of the first examples of such a grand unified theory (GUT) was developed by Glashow and Georgi in 1974.[*] One consequence of the higher symmetry required by this theory is that all elementary particles simply become facets of one another. Alan Guth confirmed to his own satisfaction that among the new particles predicted by GUTs was the *magnetic monopole*, a hypothetical single unit of magnetic 'charge' equivalent to an isolated north or south pole. In May 1979 he had begun work with a fellow postdoc, Chinese-American Henry Tye, to determine the number of magnetic monopoles likely to have been produced in the big bang. Their mission was to explain why, if magnetic monopoles were indeed formed in the early universe, none are visible today.

Guth and Tye realized that they could suppress the formation of monopoles by changing the nature of the transition from grand unified to electro-weak epochs. The monopoles largely disappeared if, instead of a smooth phase transition or 'crystallization' at the transition temperature, the universe had instead undergone *supercooling*. In this scenario, the universe persists in its grand unified state well below the transition temperature.[**]

When in December 1979 Guth explored the wider effects of the onset of supercooling, he discovered that it predicted a period of extraordinary exponential expansion of spacetime. Initially rather nonplussed by this result, he quickly realized that this explosive expansion could explain important features of the observable universe, in ways that the prevailing big bang cosmology could not.

When ice melts, the symmetry is increased and the water molecules are freed from their prison in the crystal lattice. A considerable amount of energy in the form of heat is required to complete this phase transition. This heat is used to free the water molecules from their restraints but it does not change the temperature during the transition (which remains fixed at 0°C). The transition is said to be *endothermic* – it requires the input of thermal energy. Conversely, when water freezes

---

[*] Although 'grand' and 'unified', GUTs do not seek to include the force of gravity. Theories that do so are often referred to as Theories of Everything, or TOEs.

[**] Liquid water can be supercooled to temperatures up to 40 degrees below freezing.

to form ice, this energy or 'latent heat' is released and the transition is said to be *exothermic* – there is an output of thermal energy.

In supercooled water, the water molecules remain in the liquid state well below the normal freezing point. In this situation the latent heat is not released. It remains latent.

What Guth found was that if a tiny part of the early post-big-bang universe supercooled at the end of the grand unification epoch, the energy of the Higgs field would remain similarly latent. The energy of the false vacuum so created would instead be pumped into spacetime, acting as a negative pressure and blowing spacetime up exponentially, doubling the size of the universe every ten million billion billion billionth (10^{-34}) of a second. This is Einstein's cosmological constant on acid.

'I do not remember ever trying to invent a name for this extraordinary phenomenon of exponential expansion,' Guth later wrote, 'but my diary shows that by the end of December I had begun to call it *inflation*.'[13]

Eventually, the Higgs field does crystallize and small bubbles of true vacuum start to appear, which cluster together and coalesce. The energy of the Higgs field is released, reheating the now greatly expanded universe, and channelling into the formation of a plasma of quarks, electrons, neutrinos and (mostly) radiation.

At a stroke, Guth solved two of the most stubborn paradoxes of big bang cosmology. If indeed a tiny part of the universe had supercooled and inflated in this way, then a flat, isotropic, homogeneous visible universe filled with CMB radiation with a uniform temperature was no longer a mystery. It was to be *expected*.

Whatever the local curvature of space in that part of the early universe that inflated, inflation would have flattened it. The skin of a deflated balloon may be wrinkled like a prune, curving this way and that. Pump it up and the wrinkles are quickly flattened out. The visible universe is isotropic and homogeneous because it comes from a tiny bit of grand-unified universe blown up out of all proportion by inflation. The temperature of the CMB radiation is uniform because, at the onset of inflation, all the contents of the grapefruit-sized universe were in causal contact.

COBE and WMAP

Inflationary cosmology underwent some modifications largely as a result of further tinkering with the properties of the Higgs fields used to break the symmetry of the universe at the end of the grand unified epoch. Experimental results gained in the early 1980s provided no support for early GUTs of the Georgi–Glashow type. No longer constrained by theories derived from particle physics, cosmologists were free to fit the observable universe by further tweaking the Higgs fields, which became collectively known as the *inflaton* field to emphasize its significance.

Cosmologists began to realize that the seeds of the observable structure of stars, galaxies and clusters of galaxies had to come from quantum fluctuations in the early universe, amplified by inflation.

Evidence for primeval quantum fluctuations in the inflaton field could be found in the form of tiny variations in the temperature of the CMB radiation. On 18 November 1989, the Cosmic Background Explorer (COBE) satellite was placed into sun-synchronous orbit.*

The theoretical predictions were borne out in spectacular fashion on 23 April 1992, when a team of American scientists announced that they had found these primeval fluctuations in the data from COBE. The announcement was reported worldwide as a fundamental scientific discovery and ran on the front page of the *New York Times*.

These tiny temperature variations have since been measured in even more exquisite detail by the Wilkinson Microwave Anisotropy Probe (WMAP), which was launched on 30 June 2001 and is 45 times more sensitive than COBE. It was originally intended that WMAP would provide observations of the CMB radiation for two years, but mission extensions were subsequently granted in 2002, 2004, 2006 and 2008. WMAP has produced four data releases, in February 2003, March 2006, February 2008 and January 2010.

Figure 3 shows the map of temperature variations of the CMB radiation across the sky, based on the WMAP seven-year results. In this map, interference from our own Milky Way galaxy has been carefully subtracted out. Different temperatures are shown as different greyscale

---

* It was originally planned to be launched in 1988 on a space shuttle mission, but the shuttles were grounded following the *Challenger* disaster on 28 January 1986.

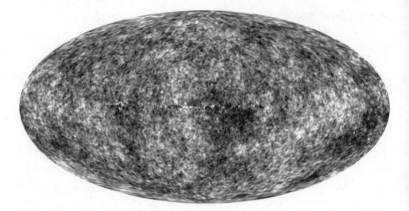

Figure 3 A detailed, all-sky image of temperature variations in the cosmic microwave background radiation based on 7-year WMAP data. The temperature variations are of the order of ± 200 millionths of a degree and are shown as greyscale colour differences. They represent quantum fluctuations that became 'frozen' in the background radiation when this disengaged from matter about 380,000 years after the big bang and acted as seeds for subsequent galaxy formation. NASA/WMAP Science Team.

colours, with white representing a higher temperature and black representing a lower temperature. The total temperature variation from white to black is just 200 millionths of a degree.

If the temperature variations from point to point appear random in this picture, it is because this is precisely what they are. The variations are due to quantum fluctuations in the inflaton field predicted, to an accuracy of one part in 10,000, by the theory of inflation.

From their vantage points in earth orbit, COBE and WMAP have detected the signature of quantum uncertainty in the birth of our universe, a signature written 13.7 billion years ago.

Dark matter

Inflation resolved some of the paradoxes of big bang cosmology, but it came with a significant trade-off. It seems that only a tiny part of the universe inflated at the end of the grand unified epoch to form the flat universe we now inhabit. We therefore have to reconcile ourselves to the fact that our observations are limited to this visible universe, our own local bubble of inflated spacetime. There may be much, much more to the larger universe, perhaps consisting of many other bubbles. We will likely never know.

This is a sobering thought, but we shouldn't dwell on it too long. We may be limited in terms of the scope of our vision, but there's still much to do.

The Coma cluster is a large galactic cluster located in the constellation Coma Berenices. It contains over a thousand identified galaxies. In 1934, the Swiss astronomer Fritz Zwicky estimated the cluster's total mass from observations of the motions of galaxies near its edge. He then compared this to another estimate based on the number of observable galaxies and the total brightness of the cluster. He was shocked to find that these estimates differed by a factor of 400.[*] The mass, and hence the gravity, of the visible galaxies in the cluster is far too small to explain the speed of the orbits of galaxies at the edge.

As much as 90 per cent of the mass required to explain the size of the gravitational effects appeared to be 'missing', or invisible. Zwicky inferred that there must be some invisible form of matter which makes up the difference. He called it 'missing mass'.

Missing mass was acknowledged as a problem but lay relatively dormant for another forty years. In 1975, the same problem was identified in the context of the motions of individual galaxies by young American astronomer Vera Rubin and her colleague Kent Ford. They reported the results of meticulous measurements of the speeds of stars at the edges of spiral galaxies. What they discovered was that stars at the edge of a galaxy move much faster than predicted based on estimates of the mass of the galaxy derived from its visible stars.

Simple Newtonian gravity predicts that the further the star is from the centre of the galaxy where most of the stellar mass is concentrated, the weaker the force of gravity that drags it around. Consequently, the speeds of stars being dragged around the centre would be expected to fall the further they are from the centre. Rubin and Ford found that the speeds of the stars actually level off the further they are from the centre.

Although Rubin's results were initially greeted with scepticism, they were eventually accepted as correct. In fact, Peebles and his

---

[*] The discrepancy was reduced by subsequent analysis, but it remains significant.

Princeton colleague Jeremiah Ostriker had earlier concluded that the observed motions of galaxies (including our own Milky Way) could not be modelled theoretically unless it was assumed that each galaxy possesses a large halo of invisible matter, effectively doubling the mass. In 1973 they had written: '... the halo masses of our Galaxy and of other spiral galaxies *exterior* to the observed disks may be extremely large'.[14] The following year they suggested that the masses of galaxies might have been underestimated by a factor of ten or more.

The missing matter was now 'dark matter'.

It is now thought possible to account for a small proportion of dark matter using exotic astronomical objects composed of ordinary matter which emit little or no radiation (and are therefore 'dark'). These are called Massive Astrophysical Compact Halo Objects, or MACHOs. Candidates include black holes or neutron stars as well as brown dwarf stars or unassociated planets.

However, the vast majority of the dark matter is thought to be so-called 'non-baryonic' matter, i.e. matter that involves not protons or neutrons, but most likely particles not currently known to the standard model of particle physics. Such particles are called Weakly Interacting Massive Particles, or WIMPs.* They have many of the properties of neutrinos, but are required to be far more massive and therefore move much more slowly.

Dark energy and the return of the cosmological constant

In 1992, the COBE satellite revealed the imprint of quantum fluctuations on the early universe and confirmed, with some confidence, that the universe is flat. Cosmic inflation could explain why this must be so.

But a density parameter Ω equal to 1 *demands* that the universe should contain the critical density of mass–energy. The problem was that, no matter how hard they looked, astronomers couldn't find it. Even dark matter didn't help. In the 1990s, the best estimate for Ω based on the observed (and implied) mass–energy of the universe was

---

* Of course, these acronyms are not coincidental. WIMP was coined first, apparently inspiring the subsequent development of MACHO.

of the order of 0.2. If this were really the case, the universe should be 'open'. With insufficient mass-energy to halt expansion, it would be destined to expand for ever.

Some astrophysicists began to mutter darkly that Einstein's infamous fudge factor – the cosmological constant – might after all have a place in the equations of the universe. Several earlier attempts had been made to resurrect it, but it had stayed dead. Now some were beginning to appreciate that a mass-energy density of 0.2 and a cosmological constant contributing another 0.8 might be the only way to explain how Ω can be equal to 1 and why the universe is flat.

To predict the ultimate fate of the universe, astronomers first have to learn as much as possible about its past. To learn about the past, it is necessary to study the most distant objects visible. Light from distant objects shows us how these objects were behaving at the time the light was emitted (for example, light from the sun shows us how the sun looked about eight minutes ago; light from the Andromeda galaxy shows us how it was 2.5 million years ago).

Because the speed of light is finite, it takes some time for this light to reach us here on earth. So when we study the light from objects at the edges of the universe, we take a journey into the distant past.

The attentions of astronomers turned to supernovae, first discovered by Zwicky in the 1920s. These occur when a star exhausts its nuclear fuel and can no longer support itself against the force of gravity. The star collapses. This collapse may trigger a sudden release of gravitational energy, blowing the star's outer layers away in a massive stellar explosion. Or, under the right conditions, a so-called white dwarf star consisting mostly of carbon and oxygen may gain sufficient mass to trigger runaway carbon fusion reactions, resulting in a similar cataclysmic explosion.

For a brief time, a supernova may outshine an entire galaxy, before fading away over several weeks or months.

Astronomers classify supernovae based on the different chemical elements they are observed to contain from their absorption spectra. Supernovae involving stars that have exhausted their hydrogen are called Type I. These are further subcategorized according to the presence of spectral lines from other atoms or ions. Type Ia supernovae produce no hydrogen absorption lines but exhibit a strong line due to ionized silicon. They are produced when a white dwarf star accretes mass (perhaps from a neighbouring star) sufficient to trigger carbon fusion reactions.

Type Ia supernovae are of interest because their relatively predictable luminosity and light curve (the evolution of their luminosity with time) means they can be used as 'standard candles'. In essence, finding a Type Ia supernova and determining its peak brightness provides a measure of the distance of its host galaxy.

Galaxies that would otherwise be too dim to perceive at the edges of the visible universe are lit up briefly by the flare of a supernova explosion.

In early 1998, two independent groups of astronomers reported the results of measurements of the redshifts and hence speeds of distant galaxies that had contained Type Ia supernovae. These were the Supernova Cosmology Project (SCP), based at the Lawrence Berkeley National Laboratory near San Francisco, California, headed by American astrophysicist Saul Perlmutter; and the High-z (meaning high-redshift) Supernova Search Team formed by Australian Brian Schmidt and American Nicholas Suntzeff at the Cerro Tololo Inter-American Observatory in Chile.

The groups were rivals in the pursuit of data that would allow us to figure out the ultimate fate of the universe. Not surprisingly, their strongly competitive relationship had been fraught, characterized by bickering and disagreement. But in February 1998 they came to agree with each other – violently.

It had always been assumed that, following the big bang and a period of rapid inflation, the universe must have settled into a phase of more gentle evolution, either continuing to expand at some steady rate or winding down, with the rate of expansion slowing. However, observations of Type Ia supernovae now suggested that the expansion of the universe is actually *accelerating*.

'Our teams, especially in the US, were known for sort of squabbling a bit,' Schmidt explained at a press conference some years later. 'The accelerating universe was the first thing that our teams ever agreed on.'[15]

Adam Reiss, a member of the High-z team, subsequently found a very distant Type Ia supernova that had been serendipitously photographed by the Hubble Space Telescope during the commissioning of a sensitive infrared camera. It had a redshift consistent with a distance of 11 billion light years, but it was about twice as bright as it had any right to be. It was the first glimpse of a supernova that had been triggered in a period when the expansion of the universe had been decelerating.

The result suggested that about five billion years ago, the expansion had 'flipped'. As expected, gravity had slowed the rate of expansion of the post-big-bang universe, until it reached a point at which it had begun to accelerate again. And there was really only one thing that could do this.

The cosmological constant was back.

One supernova does not a summer make, but in 2002, astronauts from the Space Shuttle *Columbia* installed the Advanced Camera for Surveys (ACS) on the Hubble Space Telescope. Reiss, now heading the Higher-z Supernova Search Team, used the ACS to observe a further 23 distant Type Ia supernovae. The results were unambiguous. The accelerating expansion has since been confirmed by other astronomical observations and further measurements of the CMB radiation.

There are other potential explanations, but a consensus has formed around a cosmological constant with a value of 0.73. Its reintroduction into the gravitational field equations is thought to belie the existence of a form of vacuum energy – the energy of 'empty' spacetime – which acts to push spacetime apart, just as Einstein had originally intended.

We have no idea what this energy is, and in a grand failure of imagination, it is called simply 'dark energy'.

The standard model of big bang cosmology

The ΛCDM model is based on six parameters. Three of these are the density of dark energy, which is related to the size of the cosmological constant, Λ; the density of cold dark matter; and the density of baryonic matter, the stuff of gas clouds, stars, planets and us.[*]

When applied to the seven-year WMAP results, the agreement between theory and observation is remarkable. Figure 4 shows the 'power spectrum' derived from the temperature of the CMB radiation mapped by WMAP. This is a complex graph, but it is enough

---

[*] Actually, the densities of dark matter and of baryonic matter are *derived* from the model parameters.

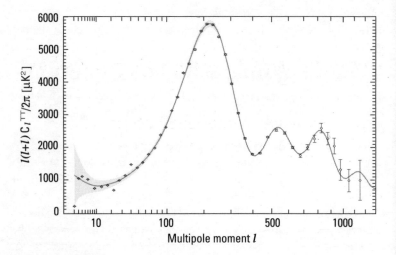

Figure 4 The 7-year temperature power spectrum from WMAP. The curve is the ΛCDM model best fit to the 7-year WMAP data. NASA/WMAP Science Team. See D. Larson et al., *The Astrophysical Journal Supplement Series, 192 (February 2011), 16.*

to know that the oscillations in this spectrum are due to the physics of the plasma that prevailed at the time of recombination. In essence, as gravity tried to pull the atomic nuclei in the plasma together, radiation pressure pushed them apart, and this competition resulted in so-called acoustic oscillations.

The positions of these oscillations and their damping with angular scale determine the parameters of the ΛCDM model with some precision. In Figure 4, the points indicate the WMAP power spectrum data with associated error bars, and the continuous line is the 'best-fit' prediction of the ΛCDM model obtained by adjusting the six parameters. In this case, the best-fit suggests that dark energy accounts for about 73.8 per cent of the universe and dark matter 22.2 per cent.

What we used to think of as 'the universe' accounts for just 4.5 per cent. This means that the evolution of the universe to date has been determined by the push-and-pull between the antigravity of dark energy and the gravity of (mostly) dark matter.

It seems that visible matter, of the kind we tend to care rather more about, has been largely carried along for the ride.

6

What's Wrong with this Picture?

Why the Authorized Version of Reality Can't be Right

The truth of a theory can never be proven, for one never knows if future experience will contradict its conclusions.

Albert Einstein[1]

The last four chapters have provided something of a whirlwind tour of our current understanding of light, matter and force, space, time and the universe. Inevitably, I've had to be a bit selective. It's not been possible to explore all the subtleties of contemporary physics, and the version of its historical development that I've provided here has been necessarily 'potted'.

I would hope that as you read through the last four chapters, you remembered to refer back to the six principles that I outlined in Chapter 1. I'd like to think that the developments in physical theory in the last century amply demonstrate the essential correctness of these principles, if not in word then at least in spirit.

Quantum theory really brings home the importance of the Reality Principle. The experimental tests of Bell's and Leggett's inequalities tell us fairly unequivocally that we can discover nothing about a reality consisting of things-in-themselves. We have to settle instead for an empirical reality of things-as-they-are-measured. This is no longer a matter for philosophical nit-picking. These experimental tests of quantum theory are respectfully suggesting that we learn to be more careful about how we think about physical reality.

I haven't been able to provide you with all the observational and experimental evidence that supports quantum theory, the standard model of particle physics, the special and general theories of relativity and the ΛCDM model of big bang cosmology. But there should be enough in here to verify the Fact Principle. It is simply not possible to make observations or perform experiments without reference to a supporting theory of some kind. Think of the search for the Higgs boson at CERN, the bending of starlight by large gravitating objects, or the analysis of the subtle temperature variations in the CMB radiation.

We have also seen enough examples of theory development to conclude that the Theory Principle is essentially correct. Abstracting from facts to theories is highly complex, intuitive and not subject to simple, universal rules. In some cases, theories have been developed in response to glaring inconsistencies between observation, experiment and the prevailing methods of explanation. Such developments have been 'data-led', with observation or experiment causing widespread bafflement before a theoretical resolution could be found.

Sometimes the theoretical resolution has been more baffling than the data, as Heisenberg himself could attest, wandering late at night in a nearby park after another long, arduous debate with Bohr in 1927. Could nature possibly be as absurd as it seemed?

In other cases, theories have been sprung almost entirely from intuition: they have been 'idea-led'. Such intuition has often been applied where a problem has been barely recognized, born from a stubborn refusal to accept inadequate explanations. Recall Einstein's light-quantum hypothesis. Remember Einstein sitting in his chair at the Patent Office, struck by the thought that if a man falls freely he will not feel his own weight.

Ideas and theories that follow from intuition can clearly precede observation and experiment. The notion that there should exist a meson formed from a charm and anti-charm quark preceded the discovery of the J/ψ in the November revolution of 1974. The Higgs mechanism of electro-weak symmetry-breaking preceded the discovery of weak neutral currents, the W and Z particles and (as seems likely) the Higgs boson.

I've tried to ensure that my descriptions of the theoretical structures that make up the authorized version of reality have been liberally sprinkled with references to the observations and experiments that have provided critically important tests. Although there are inevitable grey areas, in general the theories that constitute the authorized version are regarded as testable, and have been rigorously tested to a large degree. Perhaps you are therefore ready to accept the Testability Principle.

Then we come to the Veracity Principle. It might come as a bit of a shock to discover that scientific truth is transient. What is accepted as true today might not be true tomorrow. But look back at how the 'truth' of our universe has changed from Newton's time, or even over the last thirty years. Or even within the short period in which the Higgs boson took a big step towards becoming a 'real' entity.

Finally, there is the Copernican Principle. Nowhere in the authorized version will you find any reference to 'us' as special or privileged observers of the universe. As we currently understand it, the physics of this version of reality operates without intention and without passion. We are just passively carried along for the ride.

Now, you might have got the impression from the last four chapters that the authorized version of reality is a triumph of the human intellect and, as such, pretty rock-solid, possibly destined to last for all time. The four theoretical structures that make up the authorized version undoubtedly represent the pinnacle of scientific achievement. We should be – and are – immensely proud of them.

But these theories are riddled with problems, paradoxes, conundrums, contradictions and incompatibilities. In one sense, they don't make sense at all.

They are not the end. The purpose of this chapter is to explain why, despite appearances, the authorized version of reality can't possibly be right.

Some of these problems were hinted at in previous chapters, but here we will explore them in detail. It is important to understand where they come from and what they imply. The attempt to solve them without guidance from observation or experiment is what has led to the creation of fairy-tale physics.

The paradox of Schrödinger's cat

Actually, the problem of quantum measurement is a perfect problem for these economically depressed times. This is because it is, in fact, *three* problems for the price of one: the problem of quantum probability, the collapse of the wavefunction and the 'spooky' action-at-a-distance that this implies. Bargain!

Our discomfort begins with the interpretation of quantum probability. Quantum particles possess the property of phase which in our empirical world of experience scales up to give us wave-like behaviour. We identify the amplitude of the quantum wavefunction (or, more correctly, the modulus-square of the amplitude) as a measure of probability, and this allows us to make the connection with particles.[2] Thus, an electron orbiting a proton in a hydrogen atom might have a spherically symmetric wavefunction, and the modulus square of the amplitude at any point within the orbit relates the probability that the electron will be 'found' there.

The trouble is that phases (waves) can be added or subtracted in ways that self-contained particles cannot. We can create superpositions of waves. Waves can interfere. Waves are extended, with amplitudes in many different places. The probabilities that connect us with particles are therefore subject to 'spooky' wave effects. We conclude that one particle can have probabilities for being in many different places (although thankfully it can't have a unit or 100 per cent probability for being in more than one place at a time).

Consider a quantum system on which we perform a measurement with two possible outcomes, say 'up' or 'down' for simplicity. The accepted approach is to form a wavefunction which is a superposition of both possible outcomes, including any interference terms. This represents the state of the system prior to measurement. The measurement then collapses the wavefunction and we get a result – 'up' or 'down' with a probability related to the modulus-squares of the amplitudes of the components in the superposition.

Just *how* is this meant to work?

The collapse of the wavefunction is essential to our understanding of the relationship between the quantum world and our classical world of experience, yet it must be added to the theory as an ad hoc assumption. It also leaves us pondering. Precisely *where* in the causally connected chain of events from quantum system to human perception is the collapse supposed to occur?

Inspired by some lively correspondence with Einstein through the summer of 1935, Austrian physicist Erwin Schrödinger was led to formulate one of the most famous paradoxes of quantum theory, designed to highlight the simple fact that the theory contains no prescription for precisely how the collapse of the wavefunction is meant to be applied or where it is meant to occur. This is, of course, the famous paradox of Schrödinger's cat.

He described the paradox in a letter to Einstein as follows:

> Contained in a steel chamber is a Geiger counter prepared with a tiny amount of uranium, so small that in the next hour it is just as probable to expect one atomic decay as none. An amplified relay provides that the first atomic decay shatters a small bottle of prussic acid. This and – cruelly – a cat is also trapped in the steel chamber. According to the [wave]function for the total system, after an hour, *sit venia verbo*, the living and dead cat are smeared out in equal measure.[3]

Prior to actually measuring the disintegration in the Geiger counter, the wavefunction of the uranium atom is expressed as a superposition of the possible measurement outcomes, in this case a superposition of the wavefunctions of the intact atom and of the disintegrated atom. Our instinct might be to conclude that the wavefunction collapses when the Geiger counter triggers. But why? After all, there is nothing in the structure of quantum theory itself to indicate this.

Why not simply assume that the wavefunction evolves into that of a superposition of the wavefunctions of the triggered and untriggered Geiger counter? And why not further assume that this evolves too, eventually to form a superposition of the wavefunctions of the live and dead cat? This is what Schrödinger meant when he wrote about the living and dead cat being 'smeared out' in equal measure.

We appear to be trapped in an infinite regress. We can perform a measurement on the cat by lifting the lid of the steel chamber and ascertaining its physical state. Do we suppose that, at that point, the wavefunction collapses and we record the observation that the cat is alive or dead as appropriate?

On the surface, it really seems as though we ought to be able to resolve this paradox with ease. But we can't. There is obviously no evidence for peculiar superposition states of live-and-dead things or of 'classical' macroscopic objects of any description.* We can avoid the infinite regress if we treat the measuring instrument (in this case, the Geiger counter) as a classical object and argue that classical objects cannot form superpositions in the way that quantum objects can.

But the questions remain: why should this be and how does it work? Perhaps more worryingly, if some kind of external 'classical' macroscopic measuring device is required, then precisely what was it that, in the early moments of the big bang when the universe was the size of a quantum object, collapsed the wavefunction of the universe? Is it necessary for us to invoke an ultimate 'measurer-of-all-things'?

Irish theorist John Bell called this seemingly arbitrary split between measured quantum object and classical perceiving subject 'shifty':

> What exactly qualifies some physical systems to play the role of 'measurer'? Was the wavefunction of the world waiting to jump for thousands of years until a single-celled living creature appeared? Or did it have to wait a little longer, for some better qualified system ... with a PhD?[4]

When the wavefunction collapses, it does so *instantaneously*. That doesn't seem like much of a problem if you say it quickly and move on to something else. But, of course, this is a *big* problem. In our universe nothing, but nothing, happens instantaneously across large distances. If I happen to make a measurement on one of a pair of entangled photons that has travelled halfway across the universe before reaching my detector, how does the photon on the other side of the universe discover what result I got? Surely this is completely at odds with Einstein's special theory of relativity, which assumes that the speed of light cannot be exceeded?

This isn't a hypothetical scenario. Experiments have been performed on entangled photons which have allowed a lower limit to be placed on the speed with which the wavefunction had to collapse to give the

---

* But it's worth remembering that under certain circumstances it is possible to form superpositions of macroscopic dimensions, such as (for example) superpositions of a couple of billion electrons travelling *in opposite directions* around a superconducting ring with a diameter over a hundred millionths of the metre. Okay, these aren't cat-sized dimensions, but precisely where are we supposed to draw the line?

results observed. This speed was estimated to be at least twenty thousand times faster than the speed of light.

Many physicists have reconciled themselves to this kind of result by noting that, despite the apparent speed of the collapse process (if it really happens at all), it cannot be used to communicate information. No matter how hard we try, we cannot take advantage of this seeming example of an instantaneous physical effect to send messages of any kind. And this, they claim, allows quantum measurement peacefully (if rather uneasily) to co-exist with special relativity.

Experiments designed to test the foundations of quantum theory, such as the tests of Bell's and Leggett's inequalities, have all served to confirm the essential correctness of the theory. But they also serve to deepen the mystery. Instead of being reassured, many physicists have become increasingly alarmed, as Bell himself noted in 1985 of the Aspect experiment:

> It is a very important experiment, and perhaps it marks the point where one should stop and think for a time, but I certainly hope it is not the end. I think that the probing of what quantum mechanics means must continue, and in fact it will continue, whether we agree or not that it is worth while, because many people are sufficiently fascinated and perturbed by this that it will go on.[5]

The quantum measurement problem is extremely stubborn. We should be clear that *any* theoretical structure which attempts to resolve it takes us beyond quantum theory and therefore beyond the current authorized version of reality. Some of these attempts have led to some really bizarre interpretations of the physical world and, perhaps inevitably, to fairy-tale physics.

No rhyme or reason

In our rush to examine the theories that are supposed to help us make sense of the world, we didn't really stop to think about what kind of ultimate, all-encompassing theory we might actually *like* to have. I appreciate that our personal preferences or desires have no influence on nature itself, and that, to a large extent, any kind of ultimate theory must be shaped to fit nature's particular foibles. But there is nevertheless an important sense in which we seek a theory that we would find *satisfying*. And what is satisfaction but a consequence of fulfilling our personal preferences or desires?

I don't think many scientists (or non-scientists, for that matter) would dispute that a satisfying theory is one in which we are obliged to assume little or (even better) no a priori knowledge. The term a priori means in this context 'independent of experience'. In other words, the ultimate theory would be one that encapsulates all the relevant laws of physics in a consistent framework, such that all we would need to do to calculate what happens in a physical system in a specific set of circumstances would be to define the circumstances appropriately and press the 'enter' key.

To be fair, we would probably need to specify some fundamental physical constants – such as Planck's constant, the charge on the electron, the speed of light, and so on – but that would be it. Everything else would flow purely from the physics.[*]

As Einstein himself put it:

> It can scarcely be denied that the supreme goal of all theory is to make the irreducible basic elements as simple and as few in number as possible without having to surrender the adequate representation of a single datum of experience.[6]

If this is indeed the kind of vision we have in the backs of our minds, then the current standard model of particle physics, in the form QCD × QFD × QED, falls way short. It doesn't accommodate gravity. It requires a collection of 61 'elementary' particles.[**] And it is held together by a set of parameters that must be entered a posteriori (by reference to experience, meaning that they can't be calculated and so must be measured). American physicist Leon Lederman summarized the situation in 1993:

---

[*] In principle, these fundamental physical constants simply 'map' the physics to our human, terrestrial (and arbitrary) standards of observation and measurement.

[**] Count them. There are three generations each consisting of two leptons and two flavours of quark which come in three different colours (making 24 in total), the anti-particles of all these (making 48), twelve force particles – a photon, $W^\pm$ and Z^0 and eight gluons (making 60) – and a Higgs boson (61).

The idea is that twenty or so numbers must be specified in order to begin the universe. What are these numbers (or parameters, as they are called in the physics world)? Well, we need twelve numbers to specify the masses of the quarks and leptons. We need three numbers to specify the strengths of the forces ... We need some numbers to show how one force relates to another. Then we need ... a mass for the Higgs particle, and a few other handy items.[7]

Now we believe that the elementary particles acquire their mass through interactions with the Higgs field. This seems to suggest that there might be a way of calculating the masses a priori. In truth, however, we know nothing about the strengths of these interactions, other than the fact that they must produce the masses we observe experimentally. The interaction strengths cannot be deduced from within the standard model. Instead of putting the masses of the elementary particles into the standard model 'by hand', we put in the strengths of the interactions of these particles with the Higgs field necessary to reproduce the particle masses.

What kind of fundamental theory of particle physics is it that can't predict the masses of its constituent elementary particles? Answer: one that is not very satisfying.

No clue can be gained from the masses themselves. These are summarized in Figure 5, based on mass data from the Particle Data Group. For convenience, I've expressed the masses as multiples of the proton mass. Thus, the electron mass is 0.00054 times the proton mass. The top quark is 184 times the proton mass, and so on.

For a long time the various flavours of neutrino (electron, muon and tau) were thought to be massless, but evidence emerged in the late 1990s that neutrinos can change their flavour, in a process called 'oscillation'. Neutrino oscillation is not possible unless the particles possess very small masses, currently too small to measure accurately. The Particle Data Group suggests an upper limit of just 2 electron volts. For different reasons it has also proved difficult to get a handle on the masses of the up and down quarks, and ranges are therefore quoted.

So, can you spot the pattern? Of course, masses increase in each successive generation of particles: the tau is about 17 times heavier than the muon, which is about 209 times heavier than the electron. The top

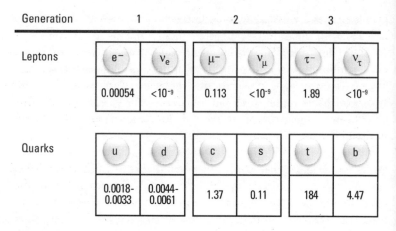

Generation	1		2		3	
Leptons	e^-	ν_e	μ^-	ν_μ	τ^-	ν_τ
	0.00054	$<10^{-9}$	0.113	$<10^{-9}$	1.89	$<10^{-9}$
Quarks	u	d	c	s	t	b
	0.0018-0.0033	0.0044-0.0061	1.37	0.11	184	4.47

Figure 5 The masses of the standard model matter particles, measured relative to the proton mass (938.27 MeV). Data are taken from listings provided by the Particle Data Group: http//pdg.lbl.gov/index.html

quark is about 134 times heavier than the charm quark, which is about 527 times heavier than the up quark. The bottom quark is about 41 times heavier than the strange quark, which is about 21 times heavier than the down quark. Three down quarks (charge –1) are about 29 times heavier than the electron (charge –1); three strange quarks are 2.9 times heavier than the muon and three bottom quarks are 7.1 times heavier than the tau.

You can keep looking at different combinations if you like, but there is no evidence for a pattern. No rhyme or reason.

Nature is probably telling us that expecting a pattern betrays our hopeless naivety. But, having found a rather consistent and compelling pattern in three generations of matter particles, I don't think it was really asking too much to expect a similar pattern in the particle masses. But there is none. And there are no clues.

Trouble with the hierarchy

The hierarchy in question here concerns the relative strengths and characteristic mass–energy scales of the fundamental forces. Specifically, in the context of particle physics it concerns the relationship between gravity and the weak force and electromagnetism: why is gravity so much weaker?

The standard model is based on the idea that the weak force and electromagnetism were once indistinguishable components of a single electro-weak force. The distinction between them was forced by symmetry-breaking at the end of the electro-weak epoch. The conditions of this period are re-created in the proton–proton collisions in the LHC, at mass-energies around a trillion (10^{12}) electron volts.

Playing the same kind of game, we suppose that the strong nuclear force and electro-weak force were once likewise indistinguishable components of a single electro-nuclear force. But to re-create the conditions characteristic of the grand unified epoch, we would need to reach mass-energies of the order of a trillion trillion (10^{24}) electron volts. Pushing even further, to conditions in which gravity merges with the electro-nuclear force to produce the single primordial force that dominated the early stages of the big bang, takes us to the Planck epoch, characterized by the *Planck mass*, about 10,000 trillion trillion (10^{28}) electron volts.[8]

Why is this a problem, exactly? Well, imposing a distinction between nature's forces as a result of breaking the prevailing symmetries is not in itself a problem. The problem is that before the symmetry is broken, all the particles involved are identical. After the symmetry is broken, the particles involved exhibit incredibly divergent masses – different by 15 orders of magnitude – consistent with the different mass-energy scales of the forces. It's as though the antenatal ultrasound reveals two monozygotic (identical twin) embryos, and yet when the symmetry is broken the mother gives birth to Arnold Schwarzenegger and Danny DeVito.

Breaking the symmetry to produce such divergent mass-energy scales appears to require an awful lot of fine-tuning. And physicists get a bit twitchy when confronted by too much coincidence.

The chickens all come home to roost when we consider the mass of the electro-weak Higgs boson. The standard quantum-theoretical approach to calculating this mass involves computing so-called radiative corrections to the particle's 'bare' mass, thereby renormalizing it. This is a calculation that has proved to be beyond the theorists, which is why nobody knew what the mass of the Higgs boson should be before the search for it began. However, it is possible to anticipate how the calculation would look in principle.

The radiative corrections involve taking account of all the different processes that a Higgs particle can undergo as it moves from place to place. These include virtual processes, involving the production of other particles and their anti-particles for a short time before these recombine back to the Higgs. The Higgs boson is obliged to couple to other particles in direct proportion to their masses, so virtual processes involving heavy particles such as the top quark would be expected to make significant contributions to the calculated mass of the Higgs.

To cut a long story short, the mass of the Higgs would be expected to mushroom as a result of these corrections. Calculations predict a mass for the Higgs that is as big as the Planck mass.

If this were really the case, then the universe would be a very different place, and neither you nor I would be around to puzzle over it. If, as seems very likely, the Higgs boson has a mass around 125 GeV, then *something* must be happening to cancel out the contributions from all those radiative corrections, and so fine-tuning the scale of the weak force.

But whatever it is, it is not to be found in the standard model.

A Prayer for Owen Meany

In John Irving's 1989 novel *A Prayer for Owen Meany*, the title character believes he is an instrument of God, and taunts his best friend John Wheelwright as they practise a basketball shot over and over again. As dusk settles over their New Hampshire playground in late November or early December 1964, first the basket and then a nearby statue of Mary Magdalene slowly disappear into the darkness.

'YOU CAN'T SEE HER, BUT YOU KNOW SHE'S STILL THERE – RIGHT?' asks Meany, in his high-pitched, childlike voice.

Yes, John knows the statue is still there.

'YOU HAVE NO DOUBT SHE'S THERE?' Meany nags.

'Of course I have no doubt,' John replies, getting exasperated.

'BUT YOU CAN'T *SEE* HER – YOU COULD BE WRONG.'

'No, I'm *not* wrong – she's there, I *know* she's there,' John yells.

'YOU ABSOLUTELY KNOW SHE'S THERE – EVEN THOUGH YOU CAN'T SEE HER?'

'Yes!'

'WELL, NOW YOU KNOW HOW I FEEL ABOUT GOD,' said Owen Meany. 'I CAN'T SEE HIM, BUT I ABSOLUTELY KNOW HE IS THERE.'[9]

This neatly summarizes our attitude to dark matter. We can't see it but we absolutely know it is there.

Dark matter is unknown to the standard model of particle physics and is therefore by definition a big problem. Explaining it is definitely going to require something beyond current theories – it is going to require 'new physics'.

Ordinary 'baryonic' matter – the stuff of everyday substances – reveals itself through its interaction with electromagnetic radiation. So, we require a form of matter that exerts a gravitational pull but is unaffected by the electromagnetic force. Several alternative varieties of dark matter have been postulated, but cold dark matter consisting of weakly interacting, slow-moving particles is thought to be most compatible with the visible structures of galaxies and galactic clusters.

There are several candidates, each more exotic than the last. WIMPs interact only via gravity and the weak force and so have many of the properties of neutrinos, except for the simple fact that they must be much heavier.

But WIMPs are not the only dark matter candidates. In 1977, Italian physicist Roberto Peccei and Australian-born Helen Quinn proposed a solution to a niggling problem in QCD, in which a hypothetical new symmetry (called the Peccei–Quinn symmetry) is spontaneously broken. Steven Weinberg and Frank Wilczek subsequently showed that this symmetry-breaking would .produce a new low-mass, electrically neutral particle, which they called the *axion*.

Although of low mass, axions can account for dark matter provided there are enough of them. Once again, calculations seemed to suggest that, if they exist, axions would have been produced in abundance in the early universe and would be prevalent today.

The Axion Dark Matter Experiment (ADMX), at the University of Washington's Center for Experimental Nuclear Physics and Astrophysics, is looking for evidence of dark matter axions interacting with the strong magnetic field generated by a superconducting magnet. In such interactions, the axions are predicted to decay into microwave photons, which can be detected.

Experiments to date have served to exclude one kind of strongly interacting axion in the mass range 1.9–3.5 millionths of an electron volt. In the next ten years the collaboration hopes either to find or exclude a weakly interacting axion in the mass range 2–20 millionths of an electron volt.

The last dark matter candidate we will consider here is the primordial black hole. Most readers will already know that a black hole is formed when a large star collapses. Its mass becomes so concentrated that not even light can escape the pull of its gravity. Astronomers have inferred the existence of two types of black hole, those with a mass around ten times the mass of the sun ($10\ M_\odot$) and super-massive black holes with masses between a million and ten billion $M_\odot$ that reside at the centres of galaxies.

Primordial black holes are not formed this way. It is thought that they might have been created in the early moments of the big bang, when wild fluctuations in the density of matter might have tripped over the threshold for black hole formation. Unlike black holes formed by the collapse of stars, primordial black holes would be small, with masses similar to those of asteroids, or about a ten billionth of $M_\odot$. To all intents and purposes, they would behave like massive particles.

Options for detecting them are limited, however. In 1974, Stephen Hawking published a paper suggesting that, contrary to prevailing opinion, large black holes might actually *emit* radiation as a result of quantum fluctuations at the black hole's event horizon, the point of no return beyond which nothing – matter or light – can escape. This came to be known as Hawking radiation. Its emission causes the black hole to lose mass and eventually evaporate in a small explosion (at least, small by astronomical standards).

If primordial black holes were formed in the early universe, and if they emit Hawking radiation, then we may be able to identify them through telltale explosions in the dark matter halo surrounding our own Milky Way galaxy. One of the many tasks of the Fermi Gamma-ray Space Telescope, launched by NASA on 11 June 2008, is to look for exploding primordial black holes based on their expected 'signature' bursts of gamma rays.

So, lots of ideas and lots of searching. But no evidence one way or another, yet.

The catastrophe of the vacuum

We saw in the last chapter that a series of astronomical observations are now lined up behind the existence of dark energy, manifested in the ΛCDM model of big bang cosmology as a cosmological constant contributing an energy density equivalent to Ω of 0.73.

At first, this seems quite baffling. It requires 'empty' spacetime to possess energy (the eponymous 'dark' energy) which acts like a kind of antigravity, pushing space apart and accelerating the expansion of the universe.

But wait. Haven't we already come across something like this before? The operation of Heisenberg's uncertainty principle on the vacuum of 'empty' space means that it is, in fact, filled with quantum fluctuations. And the existence of these fluctuations is demonstrated by the experiment which measured the Casimir force between two parallel plates.

The existence of vacuum fluctuations means that there is what physicists call a non–zero vacuum expectation value (a kind of fancy average value) for the energy of the vacuum. This sounds perfect. Surely the cosmological constant reflects the basic quantum uncertainty of spacetime. Wouldn't that be poetic?

Not so fast.

The size of the cosmological constant required by the ΛCDM model suggests a density of vacuum energy of the order of a millionth of a billionth (10^{-15}) of a joule per cubic centimetre. Now, you may not be entirely familiar with the joule as a unit of energy, so let's try to put this figure into some sort of perspective.

Kellogg's cornflakes contain about 1.6 million joules of energy per 100 grams.* A typical box contains 450 grams and has dimensions 40 × 30 × 5 centimetres. In an upright box the cornflakes tend to shake down and occupy only a proportion – let's say 75 per cent – of the volume inside. From this information we can estimate a chemical energy density of a box of cornflakes of about 1,600 joules per cubic centimetre. This gives us a sense of the scale of energy densities in 'everyday' life.

Although we don't normally include it in our daily nutritional considerations, cornflakes also contain energy in the form of mass. We can use Einstein's formula $E = mc^2$ to calculate that a 450 gram box is equivalent to about 40 million billion joules. This gives a mass-energy density of about 7 trillion (7×10^{12}) joules per cubic centimetre. There's not much to be gained by adding the chemical energy density to this figure to give a total.

So, perhaps not altogether surprisingly, the energy density of the vacuum is about 10^{-27} times the energy density of an everyday object like a box of cornflakes. Hey, it might not be completely empty, but it's still a 'vacuum', after all.

What does quantum theory predict? Well, the calculation is a little problematic. On the one hand, the uncertainty principle appears to impose some fairly rigid restrictions on what can't happen in the quantum world. However, on the other hand, it is extraordinarily liberal regarding what *can* happen. And history has shown that when quantum theory says that something can happen in principle, then this something generally tends to happen in practice.

The trouble with the uncertainty principle is that it doesn't care about the size (in energy terms) of quantum fluctuations in the vacuum provided they happen on timescales consistent with the principle. It simply demands a trade-off between energy and time. So, a fluctuation of near-infinite energy is perfectly acceptable provided it happens in an infinitesimally short time.

---

* The nutritional information on a box of cornflakes indicates an energy content of 1,604 kJ (thousand joules) per 100 grams. This is chemical energy, released when the cornflakes are combusted or digested.

I think you can probably guess where this leads. Accumulating all the quantum fluctuations that are 'allowed' by the uncertainty principle results in an infinite vacuum energy density.

This is vaguely reminiscent of the problem we encountered in quantum electrodynamics, in which an electron interacts with its own self-generated electromagnetic field, resulting in an infinite contribution to the electron mass. That problem was resolved using the technique of renormalization, effectively subtracting infinity from infinity to give a finite, renormalized result. Unfortunately, renormalizing the vacuum is a non-trivial problem.

Theorists have made some headway by 'regularizing' the calculation. In essence, this involves applying an arbitrary cut-off, simply deleting from the equations all the terms relating to the highest-energy fluctuations. These occur with dimensions and within timescales where in any case the theorists are no longer confident about the validity of quantum theory. This is the Planck scale, the domain of gravity.

Now it's certainly true that regularizing the calculation in this way does improve things. The vacuum energy density is no longer predicted to be infinite (yay!). Instead, it's predicted to have a value of the order of 100,000 googol* (10^{105}) joules per cubic centimetre. In case you've forgotten already, the 'observed' value is 10^{-15} joules per cubic centimetre, so the theoretical prediction is out by a staggering hundred billion billion googol (10^{120}).

That's got to be the worst theoretical prediction in the history of science.

It's perhaps rather amusing to note that after all the wrangling over whether or not there *is* a cosmological constant, quantum theory is actually quite clear and unambiguous on this question. There definitely *should* be a cosmological constant – quantum theory *demands* it. But now we need to understand just how it can be so *small*.

The long and winding road to quantum gravity

Within a few short months of his final lecture on general relativity to the Prussian Academy of Sciences, Einstein was back at the Academy explaining that his new theory might need to be modified:

---

* A googol is 10^{100}. I had to look it up.

> Due to electron motion inside the atom, the latter should radiate gravitational, as well as electromagnetic energy, if only a negligible amount. Since nothing like this should happen in nature, the quantum theory should, it seems, modify not only Maxwell's electrodynamics but also the new theory of gravitation.[10]

In other words, Einstein was hinting that there should be a quantum theory of gravity.

There are a number of different ways physicists can try to construct such a theory. They can try to impose quantum rules on general relativity in a process called 'canonical quantization'. This is generally known as the canonical approach to quantum gravity. Einstein himself tended to dismiss this approach as 'childish'.

Alternatively, they can start with relativistic quantum field theory and try to make it generally covariant, meaning that the physical (quantum) laws described by the theory are independent of any arbitrary change in co-ordinate system, as demanded by general relativity. This is known as the covariant approach to quantum gravity. In a quantum field theory, gravity is described in much the same way as forces in the standard model of particle physics, in terms of the exchange of a force carrier – called the graviton – between gravitating objects.

A third approach is to start over.

Whichever approach we take, we run into a series of profound problems right at the outset. General relativity is about the motions of large-scale bodies such as planets, stars, solar systems, galaxies and the entire universe within a four-dimensional spacetime. In general relativity these motions are described by Einstein's gravitational field equations. These are complex equations because the mass they consider distorts the geometry of spacetime around it and the geometry of the spacetime around it governs how the mass moves.

But spacetime itself is contained entirely within the structure of general relativity – it is a fundamental variable of the theory. The theory itself constructs the framework within which mass moves and things happen. In this sense the theory is 'background independent': it does not presuppose the existence of a background framework to which the motions of large masses are to be referred.

Quantum theory, in contrast, presumes precisely this. It is 'background dependent', requiring an almost classical Newtonian container of space and time within which the wavefunctions of quantum particles can evolve.

We ran into problems thrown up by the uncertainty principle when we considered the energy of the vacuum. But we get even more headaches when we apply the uncertainty principle to spacetime itself. General relativity assumes that spacetime is certain: it is 'here' or 'there', curves this way or that way, at this or that rate. But the uncertainty principle *hates* certainty. It insists that we abandon this naïve notion of smooth continuity and deal instead with a spacetime twisted and tortured and riddled with bumps, lumps and tunnels – 'wormholes' connecting one part of spacetime with another.

The American physicist John Wheeler called it quantum or spacetime 'foam'. A picturesque description, perhaps, but constructing a theory on it is like trying to build on a foundation of wet sand.

There are further problems. Aside from having to confront the essentially chaotic nature of spacetime at the 'Planck scale', we also have to acknowledge that we're now dealing with distances and volumes likely to catch us out in one of the most important assumptions of conventional quantum field theory – that of point particles.

In the quantum field theories that comprise the standard model of particle physics, the elementary particles are treated as though they have no spatial extension.* All of the particle's mass, charge and any other physical property it might be carrying are assumed to be concentrated to an infinitesimally small point. This would necessarily be true of elementary particles in any quantum field theory of gravity.

Obviously, the assumption of point particles is much more likely to be valid when considering physics on scales much larger than the particles themselves. But as we start to think about physics at the dimensions of the Planck length – 1.6 hundredths of a billionth of a trillionth of a trillionth (1.6×10^{-35}) of a metre, we must begin to doubt its validity.

---

* Hang on, I hear you cry. What about the uncertainty principle? Doesn't the assumption of point-like properties for an elementary particle mean a consequent assumption of certainty in its location in space? Actually, no, it doesn't. An elementary particle like an electron may be thought of as being 'smeared' out in space because the amplitude of its wavefunction is not fixed on a single point; it is extended. But it is not the particle itself that is smeared out. What is smeared out is the probability – calculated from the modulus-square of the wavefunction – of 'finding' the point-like electron within this space.

Quantum theory and general relativity are two of the most venerated theories of physics, but, like two grumpy old men, they just don't get along. Both are wonderfully productive in helping us to understand the large-scale structure of our universe and the small-scale structure of its elementary constituents. But they are volatile and seem destined to explode whenever one is shoehorned into the other.

The physics of the very small and the physics of the very large are seemingly incompatible, even though the very large (the universe) was once very, very small. A straightforward resolution of the problem is not forthcoming. Quantum gravity lies far beyond the standard model of particle physics and general relativity, and so far beyond the current authorized version of reality.

The fine-tuning problem

When we use it to try to make sense of the world around us, science forces us to abandon our singularly human perspective. We're obliged to take the blinkers off and adopt a little humility. Surely the grand spectacle of the cosmos was not designed just to appeal to our particularly human sense of beauty? Surely the universe did not evolve baryonic matter, gas, dust, stars, galaxies and clusters of galaxies just so that *we* could evolve to gaze up in awe at it?

Of course, the history of science is littered with stories of the triumph of the rational, scientific approach over human mythology, superstition and prejudice. So, we do not inhabit the centre of the solar system, with the sun and planets revolving around the earth. The sun, in fact, is a rather unspectacular star, like many in our Milky Way galaxy of between 200 and 400 billion stars. The Milky Way is just one of about 200 billion galaxies in the observable universe. It makes no sense to imagine that this is all for our benefit.

This is the Copernican Principle.

What, then, should we expect some kind of ultimate theory of everything to tell us? I guess the assumption inherent in the scientific practice of the last few centuries is that such a theory of everything will explain *why* the universe and everything in it *has* to be the way it is. It will tell us that our very existence is an entirely natural consequence of the operation of a (hopefully) simple set of fundamental physical laws.

We might imagine a set of equations into which we plug a number of perfectly logical (and inescapable) initial conditions, then press the 'enter' key and sit back and watch as a simulation of our own universe unfolds in front of our eyes. After a respectable period, we reach a point in the simulation where the conditions are right for life.

We are not so naïve as to imagine that science will ever completely eliminate opportunities for speculation and mythologizing. There are some questions that science may never be able to answer, such as what (or who, or should that be Who?) pressed the 'enter' key to start the universe we happen to live in. But surely the purpose of science is to reduce such opportunities for myth to the barest minimum and replace them with the cold, hard workings of physical mechanism.

Here we run into what might be considered the most difficult problem of all. The current authorized version of reality consists of a marvellous collection of physical laws, governed by a set of physical constants, applied to a set of elementary particles. Together these describe how the physical mechanism is meant to work. But they don't tell us where the mechanism comes from or why it has to be the way it is.

What's more, there appears to be no leeway. If the physical laws didn't quite have the form they do have, the physical constants were to have very slightly different values, or the spectrum of elementary particles were marginally different, then the universe we observe could not exist.*

This is the fine-tuning problem.

We have already encountered some examples of fine-tuning, such as the mass-energy scales (and the relative strengths) of gravity and the weak and electromagnetic forces. More fine-tuning appears to be involved in the vacuum energy density, or the density of dark energy, responsible for the large-scale structure of the universe.

In 1999, the British cosmologist Martin Rees published a book titled *Just Six Numbers*, in which he argued that our universe exists in its observed form because of the fine-tuning of six dimensionless physical constants.

---

* And, of course, we would not exist to puzzle over what had gone wrong.

As I've already said a couple of times, physicists get a little twitchy when confronted with too many coincidences. Here, however, we're confronted not so much with coincidences but rather with conspiracy on a grand scale.

Now, we should note that some physicists have dismissed fine-tuning as a non-problem. The flatness and horizon problems in early big bang cosmology appeared to demand similarly fantastic fine-tuning but were eventually 'explained' by cosmic inflation. Isn't it the case here that we're mistaking ignorance for coincidence? In other words, the six numbers that Rees refers to are not fine-tuned at all: we're just ignorant of the physics that actually governs them.

But our continued inability to devise theories that allow us to deduce these physical constants from some logical set of 'first principles', and so explain why they have the values that they have, leaves us in a bit of a vacuum.

It also leaves the door wide open.

Clueless

There are further problems that I could have chosen to include in this chapter, but I really think we have enough to be going on with. On reading about these problems it is, perhaps, easy to conclude that we hardly know anything at all. We might quickly forget that the current authorized version of reality actually explains an awful lot of what we can see and do in our physical world.

I tend to look at it this way. Several centuries of enormously successful physical science have given us a version of reality unsurpassed in the entire history of intellectual endeavour. With a very few exceptions, it explains every observation we have ever made and every experiment we have ever devised.

But the few exceptions happen to be very big ones. And there's enough puzzle and mystery and more than enough of a sense of work-in-progress for us to be confident that this is not yet the final answer.

I think that's extremely exciting.

Now we come to the crunch. We know the current version of reality can't be right. There are some general but rather vague hints as to the directions we might take in search of solutions, but there is no flashing illuminated sign saying 'this way to the answer to all the puzzles'. And there is no single observation, no one experimental result, that helps to point the way. We are virtually clueless.

Seeking to resolve these problems necessarily leads us beyond the current version of reality, to grand unified theories, theories of everything and other, higher speculations. Without data to guide us, we have no choice but to be idea-led.

Perhaps it is inevitable that we cross a threshold.

Part II

The Grand Delusion

7

Thy Fearful Symmetry

Beyond the Standard Model: Supersymmetry and Grand Unification

Concepts are simply empty when they stop being firmly linked to experiences. They resemble social climbers who are ashamed of their origins.

Albert Einstein[1]

I guess I should now come clean. Time to own up.

When I introduced the standard model of particle physics in Chapter 3, I talked about symmetry only in the context of symmetry-breaking and the role of the Higgs field. Readers with more than a passing acquaintance with the standard model will know that this is far from the whole story.

Arguably one of the greatest discoveries in physics was made early in the twentieth century. This discovery provides us with a deep connection between critically important laws of conservation – of mass-energy, linear and angular momentum, and many other things besides – and basic symmetries in nature. And this was a discovery made not by a leading physicist, but a *mathematician*.

In 1915, German mathematician Emmy Noether deduced that the origin of the structure of physical laws describing the conservation of quantities such as energy and momentum can be traced to the behaviour of these laws in relation to certain continuous symmetry transformations.

We tend to think of symmetry in terms of transformations such as mirror reflections. In this case, a symmetry transformation is the act of reflecting an object as though in a mirror, in which left reflects right. We push this further along. We let top reflect bottom, front reflect back. We claim that an object is symmetrical if it looks the same on either side of some centre or axis of symmetry. If the object is unchanged (the technical term is 'invariant') following such an act, we say it is symmetrical.

These are examples of *discrete* symmetry transformations. They involve an instantaneous 'flipping' from one perspective to another, such as left-to-right. But the kinds of symmetry transformations identified with conservation laws in Noether's theorem are very different. They involve gradual changes, such as continuous rotation in a circle. Rotate a perfect circle through a small angle measured from its centre and the circle obviously appears unchanged. We conclude that the circle is symmetric to continuous rotational transformations. We find we can't do the same with a square. A square is not symmetric in this same sense. It is, instead, symmetric to discrete rotations through right angles.

Noether discovered that the dynamics of physical systems (the disposition of energy and momentum during a physical change) are governed by certain continuous symmetries. These symmetries are reflected in the equations – and hence the laws – describing the dynamics. The operation of these various symmetries means that the dynamical quantities that they govern are also conserved quantities. Thus, each law of conservation (of energy and momentum) is connected with a continuous symmetry transformation.

Changes in the energy of a physical system are invariant to continuous changes or 'translations' in *time*. In other words, the mathematical relationships which describe the energy of a physical system now will be exactly the same a short time later. This means that these relationships do not change with time, which is just as well. Laws that broke down from one moment to the next could hardly be considered as such. We expect such laws to be, if not eternal, then at least the same yesterday, today and tomorrow.

We can think about this another way. Suppose we discover mathematical relationships that describe the energy of a physical system and we find that these relationships do not change with time. Then we will also find that the energy so described is a quantity that is conserved – it is a quantity that can be converted from one form to another, but it cannot be created or destroyed.

A snooker player lines up a shot. He judges the speed of the cue ball and the angle of impact required to send the black into the corner pocket. He strikes the cue ball and pots the black (to ripples of applause).

His judgement and his technical skills as a snooker player are based on the principle of the conservation of linear momentum.

The momentum of an object is simply its mass multiplied by its speed in a given direction. When the cue ball strikes the black, its momentum is distributed between the cue ball and the black ball. The cue ball recoils, its mass unchanged but now with a different speed and direction. As a result of the impact, the black ball is propelled in the direction of the pocket at a certain speed. The total momentum of the cue ball and the black ball after the impact is the same as the momentum of the cue ball and the black ball prior to impact. Total momentum is conserved.

Noether found the equations describing changes in momentum to be invariant to continuous *translations in space*. The laws do not depend on location. They are the same here, there and everywhere.

An Olympic figure skater concludes her medal-winning performance. She enters a spin with arms and leg outstretched. As she draws her arms and leg back towards her centre of mass, she reduces her size as a spinning object as measured from her centre of rotation. Her linear momentum increases in compensation, and she spins faster. This is the conservation of angular momentum in action. For angular momentum, defined as motion in a circle at a constant speed, the equations are invariant to rotational symmetry transformations. They are the same irrespective of the *angle of direction* measured from the centre of the rotation.

Once the connection had been established, the logic of Noether's theorem could be turned on its head. Suppose there is a physical quantity which appears to be conserved but for which the laws governing its behaviour have yet to be worked out. If the physical quantity is indeed conserved, then the laws – whatever they are – must be invariant to a specific continuous symmetry transformation. If we can discover what this symmetry is, then we are well on the way to figuring out the mathematical form of the laws themselves.

Physicists found that they could use Noether's theorem to help them find a short cut to a theory, thus avoiding a lot of hit-and-miss speculation.

So, I can now reveal, the standard model of particle physics is built out of quantum field theories that are invariant to different continuous symmetry transformations. Consequently they respect the conservation of certain physical properties. For example, QED is invariant to symmetry transformations within something called the U(1) symmetry group. It really doesn't matter too much what this means, but we can think of the U(1) symmetry group as describing transformations synonymous with continuous rotations in a circle.[2]

Another way of representing U(1) is in terms of continuous transformations of the *phase angle* of a sine wave. Different phase angles correspond to different amplitudes of the wave in its peak–trough cycle. In the case of the electron, symmetry is preserved if changes in the phase of its wavefunction are matched by changes in its accompanying electromagnetic field. This ties the electron and the electromagnetic field together in an intimate embrace. The upshot is that, as a direct result of this phase symmetry, electric charge is conserved.

Although it is often difficult to follow the rather esoteric arguments from symmetry, make no mistake: this is powerful stuff. Electro-weak theory was constructed from an SU(2) quantum field theory 'multiplied' by the U(1) quantum field theory of electromagnetism, written SU(2) × U(1). Quantum chromodynamics is an SU(3) quantum field theory.[3] The standard model, which we discussed previously as the product QCD × QFD × QED, is more often written in physics books and journal articles in terms of the symmetry groups to which these theories refer: as SU(3) × SU(2) × U(1).

The numbers in brackets refer to the number of variables, dimensions or 'degrees of freedom' associated with the symmetry group. These are not the familiar dimensions of spacetime. They are abstract mathematical dimensions associated with the properties of the different symmetry groups.[*]

But although they are abstract, the dimensions have important consequences which are reflected in the properties and behaviours of particles and forces in our physical world. At this stage it's useful just to note that a U(1) quantum field theory describes a force carried by a single force particle, the photon. An SU(2) quantum field theory

---

[*] They should not be confused with the predictions of theories such as superstring theory (which we will examine in the next chapter), which suggest that there are more *spacetime dimensions* than the four with which we are already familiar.

describes a force carried by three force particles, the 'heavy photons' W^+, W^- and Z^0. Finally, an SU(3) quantum field theory describes a force carried by eight different force particles, the coloured gluons. The higher the symmetry, the greater the number of particles that tend to be involved.

Symmetry and the search for grand unification

The standard model is a triumph. But don't be misled. It is not a unified theory of the fundamental atomic and subatomic forces. We should interpret the terminology SU(3) × SU(2) × U(1) literally. This rather clumsy notation tells us that the standard model is actually a collection – a product – of a set of distinct quantum field theories which describe the different forces. It doesn't explain where these forces come from or why they have the strengths that they have. And it doesn't explain why the elementary particles that are acted on by and carry these forces have the masses that they possess.

The attempts that were made in the mid-1970s to construct a grand unified theory, or GUT, were based on the search for an appropriate symmetry group. The SU(3) × SU(2) × U(1) structure of the standard model would spring naturally from the *breaking* of this higher symmetry, requiring more Higgs-like fields and Higgs particles. In the context of the standard model of big bang cosmology, this is thought to have happened towards the end of the grand unified epoch, about a trillionth of a trillionth of a trillionth of a second (10^{-36} seconds) after the big bang.

In 1974, Sheldon Glashow and Howard Georgi thought they had found the symmetry group on which a grand unified theory could be constructed. It was the symmetry group SU(5).

It seems perfectly sensible to look for unification at higher and higher symmetries, and Glashow and Georgi's symmetry group SU(5) appeared to fit the bill. It has five 'dimensions'.* Three are needed to accommodate the SU(3) field theory of QCD and two are needed for the SU(2) × U(1) theory of QFD × QED. SU(5) is therefore the minimum symmetry required to fit everything in.

---

* Strictly speaking, SU(5) is 15-dimensional, but acts on a complex 5-dimensional space ...

At the time, this seemed like progress. A couple of the parameters of the standard model – those relating to the relative strengths of the subatomic forces – could now be tied to each other and so lost some of their apparent arbitrariness. But that was about as far as it went. The unified theory didn't resolve any of the fundamental problems of the standard model and had nothing to add to the story of the spontaneous breaking of the symmetry of the electro-weak force.

Opening up the number of 'dimensions' of the symmetry group also had consequences beyond simply accommodating the known forces and their particles. One rather obvious consequence of a world which becomes more and more symmetrical at higher and higher energies is that the tremendous diversity we see in nature at 'everyday' energies tends to disappear. The world becomes much more symmetrical. The result is really rather bland, with every particle having some kind of symmetrical relationship with every other. The now massless matter and force particles become virtually impossible to distinguish one from another, and the world is ruled by a single electro-nuclear force carried by massless bosons, called X bosons.

This means that in theories involving a higher symmetry, new symmetry relationships are established between the particles of the theory. These relationships persist when the symmetry is broken.

In the case of Glashow and Georgi's SU(5) GUT, these relationships are embodied through a series of particle 'multiplets'. The symmetry group demands a series of fundamental particle quintuplets (groups of five) and a series of decuplets (groups of ten). For the first-generation quarks and leptons, one quintuplet consists of the red, green and blue down quarks, the positron and the electron anti-neutrino (five particles). The decuplet consists of red, green and blue up quarks, anti-up quarks, anti-down quarks and the electron (ten particles). Similar particle multiplets are established for the second- and third-generation quarks and leptons.

The X boson (thought to acquire a mass of the order of ten million billion – 10^{16} – GeV as a consequence of breaking the SU(5) symmetry) can now mediate transitions between particles in each of these multiplets. The fact that these contain both quarks and leptons means that quarks and leptons can in principle be interconverted.

Imagine a proton consisting of red and green up quarks and a blue down quark. In the Georgi–Glashow model a green up quark in a proton can be converted into an anti-red anti-up quark, and a blue down quark can be converted into a positron. As a consequence, a proton can now be transformed into a meson formed from a red up quark and anti-red anti-up quark (actually, this is a neutral pion) and a positron. The pion goes on to decay into two photons.

'And then I realized that this made the proton, the basic building block of the atom, unstable,' Georgi said. 'At that point I became very depressed and went to bed.'[4]

Neutrons are known to be unstable due to the actions of the weak force. However, protons are observed to be rather more indomitable. Unstable neutrons in the nuclei of certain isotopes might make these isotopes radioactive, but life itself is incompatible with radioactive protons. The continued existence of life forms can be combined with the kinds of threshold levels of radioactivity that they can tolerate to deduce that the half-life of the proton must be at least a billion billion (10^{18}) years. Any less than this and life as we know it would not be possible.*

But whilst the quark–lepton transformations made possible in the Georgi–Glashow model exposed the proton to some instability, this was hardly a harbinger of doom. The X boson is thought to possess a huge mass, and the probability of proton decay is related to the fourth power of this mass.** Consequently, the model predicts a half-life for the proton of about a million trillion trillion (10^{30}) years, about a hundred million trillion (10^{20}) times the present age of the universe.

Now don't get confused. This long half-life doesn't mean that we would necessarily have to wait this long to observe a proton decaying. Radioactive decay is a spontaneous, random process thought to be driven by quantum fluctuations of the 'vacuum', and so it can in principle happen at any time. However, the long half-life does suggest that proton decays, if they occur at all, would be very, very rare.

---

* I actually can't write such words without hearing *Star Trek*'s McCoy in my head: 'It's life, Jim, but not as we know it.'

** To illustrate what this means, suppose the probability of proton decay is 1 per cent (it is actually much, much smaller than this). We re-write 1 per cent as 0.01 and raise this to the fourth power: $0.01^4 = 0.01 \times 0.01 \times 0.01 \times 0.01 = 0.00000001$, which is a million times smaller than 1 per cent.

Nevertheless, collect 10^{30} or so protons together in a big tank, and there's a chance you might catch one in the act of decaying.

The prediction led to a rush to establish vast underground laboratory experiments to do just this. Such experiments involve the study of large tanks of water.* The tanks are insulated from the effects of cosmic rays and other stray particles.

For example, the Super-Kamioke Nucleon Decay Experiments (Super-Kamiokande or Super-K) is a laboratory located in a mine about a kilometre beneath Mount Kamioke, near the city of Hida in Japan. The laboratory contains a stainless-steel tank holding 50,000 tonnes of ultra-pure water, surrounded by instruments tuned to detect the telltale signals of dying protons. There are similar laboratories in Kamataka, India, beneath the Alps in France and in mines in Minnesota and Ohio.

These experiments have concluded that the proton must have a half-life of the order of at least ten billion trillion trillion (10^{34}) years, about ten thousand times longer than the predictions of the Georgi–Glashow model.**

Georgi had good reasons to be depressed. This early example of a GUT yielded predictions for the stability of the proton that were not upheld by experiment. Other symmetries and algebras were tried, but no real solutions were forthcoming. An annual scientific conference on the subject of GUTs established in 1980 did not survive beyond 1989, as physicists turned their attentions elsewhere.

SUSY and the symmetry between fermions and bosons

Perhaps the problem is that SU(3) × SU(2) × U(1) is far from the full story, and we're reaching for a unified theory before we know quite what it is that we're supposed to be unifying. After all, the standard model is riddled with problems of its own. Instead of expecting these

---

* Each water molecule – H_2O – contains 18 protons, 16 in the oxygen atom nucleus and one in each of the two hydrogen nuclei.

** Readers concerned that such elaborate (and, no doubt, expensive) facilities appear to have been dedicated to an ultimately fruitless search should draw some comfort from the fact that they are also engaged in other experiments. For example, Super-Kamiokande is an exquisite neutrino observatory and has contributed precise measurements of neutrino oscillation.

to be solved through the development of GUTs, perhaps we should try to address these problems first and then look at how unification might then be achieved.

It was natural to continue to look to symmetry relationships for solutions to some of these problems, such as the hierarchy problem and its implications for the theoretical evaluation of the Higgs mass. Perhaps all the fine-tuning is more apparent than real. As American physicist Stephen Martin put it:

> The systematic cancellation of the dangerous contributions to [the Higgs mass – in other words, the hierarchy problem] can only be brought about by the type of conspiracy that is better known to physicists as a symmetry.[5]

The symmetry in question is called a *supersymmetry*.

We should be clear that the original motive for developing theories of supersymmetry (which we abbreviate to SUSY) was *not* driven by a compelling desire to solve all the problems of the standard model. For a hard-nosed pragmatist, demanding to know the cost of everything (and so, to quote Oscar Wilde, thereby knowing the *value* of nothing), it might be a little difficult to come to terms with the simple fact that theorists often develop theories simply because these things are interesting and form structures of great intrinsic *beauty*.

And yet this was the situation in the early 1970s, when the first theories of supersymmetry were developed by a number of Soviet physicists based in Moscow and Kharkov. The theory was independently rediscovered in 1973 by CERN physicists Julius Wess and Bruno Zumino.* As these theorists had not set out to solve the problems of the standard model, for a time SUSY was regarded (as was laser technology in the 1960s) as a solution in search of a problem. And yet it has become one of the most logical and compelling ways to transcend standard model physics.

SUSY is not a GUT. It is probably best to think of it as an important stepping stone. If it can be shown that the world is indeed

---

* As is typical, the history is a little more complicated than this, with several theorists discovering and then rediscovering the basic principles.

supersymmetric, then some (not all) of the problems of the present standard model go away and the path to a GUT is clearer. SUSY, however, is like life. It involves a big compromise. In return for some potentially neat solutions, what we get is a *shedload* of new particles.

As Noether discovered, continuous transformations of spacetime symmetries – of time, space and rotation – are associated with the conservation of energy and linear and angular momentum. An important theorem developed in the late 1960s suggested that this was it – there could be no further spacetime symmetries. However, it soon became apparent that the theorem was not watertight; it contained a big loophole. It assumed that in any symmetry transformation, fermions would continue to be fermions and bosons would continue to be bosons.

This brings us to an important assumption.

> **The supersymmetry assumption**. *In essence, SUSY is based on the assumption that there exists a fundamental spacetime symmetry between matter particles (fermions) and the force particles (bosons) that transmit forces between them, such that these particles can transform into each other.*

It is essential to our understanding of what follows that we grasp the significance of this last sentence.

The standard model of particle physics and the great variety of experimental data that has been gathered to validate it offer no real clues as to how its rather obvious failings can be addressed and corrected. Physicists have no real choice therefore but to go with their instincts.

And their instincts tell them that at the heart of the solution must lie some kind of fundamental symmetry. As Gordon Kane, a leading spokesperson for supersymmetry, explained:

Supersymmetry is the idea, or hypothesis, that the equations of [a unified theory] will remain unchanged even if fermions are replaced by bosons, and vice versa, in those equations in an appropriate way. This should be so in spite of the apparent differences between how bosons and fermions are treated in the standard model and in quantum theory ... It should be emphasized that supersymmetry is the *idea* that the laws of nature are unchanged if fermions [are interchanged with] bosons.[6]

Assumption, idea, hypothesis. Call it what you will, SUSY is basically an enormous bet. With no real clues as to how physics beyond the standard model should be developed, physicists have decided that they must take a gamble.

Okay, but if we're going to bet the farm on Lucky Boy, running in the 3.15 at Chepstow, we will typically need a damn good reason. Why pin all our hopes on a fundamental spacetime symmetry between fermions and bosons? For the simple reason that a symmetry of this kind offers exactly the kind of cancellation required to fine-tune the Higgs mass.

Stephen Martin again:

> Comparing [these equations] strongly suggests that the new symmetry ought to relate fermions and bosons, because of the relative minus sign between fermion loop and boson loop contributions to [the Higgs mass] ... Fortunately, the cancellation of all such contributions to scalar masses is not only possible, but is actually unavoidable, once we merely assume that there exists a symmetry relating fermions and bosons, called a *supersymmetry*.[7]

Superpartners and supersymmetry-breaking

In fact, supersymmetry is not so much a theory as a *property* of a certain class of theories. There are therefore many different kinds of supersymmetric theories. To keep things reasonably simple, I propose to explore some of the consequences of supersymmetry by reference to something called the Minimal Supersymmetric Standard Model, usually abbreviated as the MSSM. This was first developed in 1981 by Howard Georgi and Greek physicist Savas Dimopoulos,[*] and is the simplest supersymmetric extension to the current standard model of particle physics.

---

[*] Yet again, the historical development of the MSSM is more complicated that this simple statement implies. See http://cerncourier.com/cws/article/28388/2 for more details. Thanks to Peter Woit for bringing this article to my attention.

So, what are the consequences of assuming a symmetry relation between fermions and bosons in the MSSM? The answer is, perhaps, relatively unsurprising. In the MSSM, the various particle states form *supermultiplets*, each of which contains both fermions and bosons which are 'mirror images' of each other.

Every particle in the standard model has a *superpartner* in the MSSM. For every fermion (half-integral spin) in the standard model, there is a corresponding so-called scalar fermion, or *sfermion*, which is actually a boson (zero spin). For every boson in the standard model, there is a corresponding *bosino*, which is actually a fermion with spin ½.

To generate the names of fermion superpartners we prepend an 's' (for scalar). So, the superpartner of the electron is the scalar electron, or selectron. The muon is superpartnered by the smuon. The superpartner of the top quark is the stop squark, and so on.

To generate the names of the boson superpartners we append '-ino'. The superpartner of the photon is the photino. The W and Z particles are partnered by winos* and zinos. Gluons are partnered by gluinos. The Higgs boson is partnered by the Higgsino (actually, several Higgsinos). Once you've grasped the terminology, identifying the names and the properties of the superpartners becomes relatively straightforward. The superpartners of the standard model particles are summarized in Figure 6.

This all seems rather mad. And yet we've come across something very similar once already in the history of quantum and particle physics. There are precedents for this kind of logic. Recall that Paul Dirac discovered – from purely theoretical considerations – that for every particle there must exist an anti-particle. A negatively charged electron must be 'partnered' by a positive electron. The positron was discovered shortly after Dirac's discovery, and anti-particles are now familiar components of the standard model.

The symmetry between particle and anti-particle, between positive and negative and negative and positive, is an exact symmetry. This means that a positron has precisely the same mass as an electron and is to all intents and purposes identical to an electron but for its charge. Similarly for all other particle–anti-particle pairs.

---

* Pronounced 'weenos'. Presumably to avoid confusion.

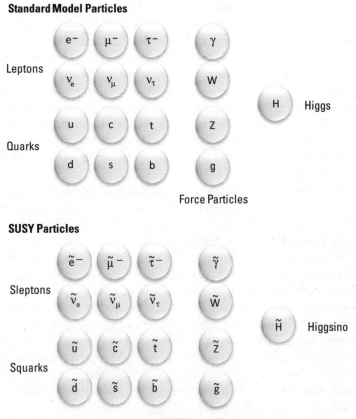

Figure 6 Supersymmetry predicts that every particle in the standard model must possess a corresponding superpartner. Matter particles – leptons and quarks (which are fermions) – are partnered by sleptons and squarks (bosons). Force particles (bosons) are partnered by bosinos (fermions), such as the photino, wino, zino and gluinos. The Higgs boson is partnered by the Higgsino.

But supersymmetry cannot be an exact symmetry. If the symmetry between fermions and bosons were exact, then we would expect the superpartners to have precisely the same masses as their standard model counterparts. In such a situation, we wouldn't be talking about assumptions, ideas or hypotheses, because our world would be filled with selectrons and massless photinos.* These super-particles (or 'sparticles') would already be part of our experience. And, consequently, they would already form part of our theoretical models.

You can guess where this is going. The fact that the world is not already filled with superpartners that impress themselves on our experience and shape our empirical reality suggests that the exact symmetry between partners and superpartners no longer prevails. Something must have happened to force a distinction between the standard model particles and their superpartners. In other words, at some time, presumably shortly after the big bang, the supersymmetry must have been broken.

We assume that the superpartners gained mass in this process, such that they are all (rather conveniently) much heavier than the more familiar particles of the standard model, and have so far stayed out of reach of terrestrial particle colliders.

Cynicism is cheap, and not altogether constructive. It's certainly true that we seem to be building assumptions on top of assumptions. *If* superpartners exist then they *must* be massive. But, to a certain extent, suggesting that there must exist – from purely theoretical considerations – massive superpartners for every particle in the standard model is no different in principle to Dirac's discovery of antimatter.

Before we go on to consider what the MSSM predicts and how it might be tested, it's useful to review the potential of the theory to resolve many of the current problems with the standard model.

SUSY and the hierarchy problem

It is already apparent that one of the most compelling arguments in favour of SUSY is that it provides a perfectly logical, natural resolution of the hierarchy problem, at least in terms of stabilizing the electro-weak Higgs mass.

Recall from the last chapter that the hierarchy problem has two principal manifestations. There is the inexplicable gulf between the energy scale of the weak force and electromagnetism and the Planck scale. And then there is the problem that quantum corrections to the

---

* White ambassardinos of morning?

Higgs mass should in principle cause it to mushroom in size all the way to the Planck mass, completely at odds with the electro-weak energy scale and recent experiments at the LHC which suggest a Higgs boson with a mass of 125 GeV.

At a stroke, SUSY eliminates the problems with the radiative corrections. The Higgs mass becomes inflated through interactions particularly with heavy virtual particles, such as virtual top quarks. The top quark is a fermion, and in the MSSM we must now include radiative corrections arising from interactions with its corresponding sfermion, the stop squark.

Now, I have very limited experience and virtually no ability as a mathematician. But what experience I do have allows me the following insight. If, when grappling with a complex set of mathematical equations, you are able to show that all the terms cancel beautifully and the answer is zero, the result is pure, unalloyed joy.

And this is what happens in SUSY. The positive contributions from radiative corrections arising from interactions with virtual particles are cancelled by negative contributions from interactions with virtual sparticles.*

The cancellation is not exact because, as I mentioned above, the symmetry between particles and sparticles cannot be exact. This is okay; the theory can tolerate some inexactness. However, force-fitting a light Higgs mass and broken supersymmetry does place some important constraints on the theory. Most importantly, the masses of many of the superpartners cannot be excessively large. If they exist, then they must possess masses of the order of a few hundred billion electron volts up to a trillion electron volts. Much heavier, and they couldn't serve the purpose of stabilizing the Higgs at the observed mass, and hence establishing the scale of the weak force and electromagnetism.

This has important implications for the testability of the theory, which I will go on to examine below.

---

* In fact, to make this mechanism work, the MSSM actually needs five Higgs particles, each with a different mass. Three of these are neutral and two carry electric charge.

SUSY and the convergence of subatomic forces

In 1974, Stephen Weinberg, Howard Georgi and Helen Quinn showed that the interaction strengths of the electromagnetic, weak and strong nuclear forces become near equal at energies around 200,000 billion GeV. The operative term here is 'near equal'. As Figure 7(a) shows, extrapolating the standard model interaction strengths up to this scale shows that they become similar, but do not converge.

We tend to assume that the subatomic forces that we experience today are the result of symmetry-breaking applied to a unified electronuclear force shortly after the big bang. If this is really the case, then it seems logical to expect that the interaction strengths of these forces should converge on the energy that prevailed at the end of the grand unified epoch. Although this is a sizeable extrapolation – we're trying to predict behaviour at energies about 14 orders of magnitude beyond our experience – the fact that the standard model interaction strengths do not converge is perhaps a sign that something is missing from the model.

The MSSM resolves this problem, too. The effects of interactions with virtual sparticles change the extrapolation such that the strengths of the forces smoothly converge on a near-single point, as shown in Figure 7(b). This is much more like it. Now the extrapolation is strongly suggestive of a time and an energy regime where a single electro-nuclear force dominated.

If we are prepared to draw conclusions from such calculations, then it does seem that something significant is missing from the standard model, and that something could be SUSY.

SUSY, the LSP and the problem of dark matter

The problem of dark matter demands a solution that lies beyond the current standard model of particle physics. By definition, dark matter must consist of a particle or particles that are not to be found among the known families of quarks, leptons and force carriers. Dark matter candidates can only be found among new particles predicted by theories that extend the standard model in some way.

It probably won't come as much of a surprise to learn that SUSY predicts the existence of particles with precisely the right kinds of properties.

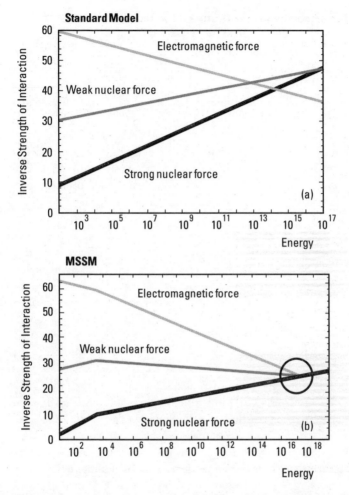

Figure 7 (a) Extrapolating the strengths of the forces in the standard model of particle physics implies an energy (and a time after the big bang) at which the forces have the same strength and are unified. However, the forces do not quite converge on a single point. (b) In the Minimum Supersymmetric Standard Model (MSSM) the additional quantum fields change this extrapolation, and the forces more nearly converge.

Remember, there is a category of candidate for cold dark matter particles known as WIMPs. These are neutrino-like (neutral, not subject to the strong force or electromagnetism) but much, much heavier. By design, the MSSM predicts superpartners that must be heavy (as they haven't been observed yet) and several of these are neutral. To take an example, the photino – the heavy superpartner of the photon – readily fits the bill.

But many potential candidate superpartners are believed to be relatively unstable. They will decay rapidly into other sparticles. We can conclude, with some sense of justification, that the cold dark matter responsible for governing the shapes, structures and rotation speeds of whole galaxies is likely to be much more enduring than this. The attentions of theorists have therefore turned to the sparticle most likely to lie at the end of the chain of decays. This is referred to as the Lightest Superpartner, or LSP. Simply put, once the sparticles have decayed into the LSP, there's nowhere else for it to go. The LSP is stable.

Now, the *neutralino* is not, as might first be thought, the superpartner of the neutrino.[*] In the MSSM, neutralinos are formed through the quantum superposition of the zino, the photino and a couple of electrically neutral Higgsinos. This mixing occurs because these particles have similar properties (such as charge and spin), and there is an unwritten law in quantum theory that if particles with mass cannot be distinguished by their other quantum properties, then they are likely to suffer an identity crisis and combine in a superposition. The mixing produces four neutralino states, the lightest of which is a candidate for the LSP and therefore a candidate WIMP.

Calculations suggest that if the MSSM is right and neutralinos do exist, then their 'relic density' – the density of neutralinos left over after the big bang – is consistent with the observed density of cold dark matter in the ΛCDM model.

Supergravity

It does seem as though we're getting quite a lot in return for our willingness to suspend disbelief and embrace the idea of heavy

---

[*] The super-partners of neutrinos are sneutrinos.

superpartners. We solve the hierarchy problem. We enable convergence between the fundamental forces that operate at atomic and subatomic levels. We gain some candidates for cold dark matter particles.

There is yet more.

In SUSY, a spacetime symmetry transformation acting on a fermion or boson changes the spin of the particle by ½. Fermions become sfermions, with spin zero. Bosons become bosinos, with spin ½. This kind of change affects the spacetime properties of the particle that is transformed, such that it is slightly displaced.* Conventional standard model forces such as electromagnetism cannot displace particles in this way. They can change the direction of motion, momentum and energy of the particle but they cannot displace it in spacetime. In fact, this displacement is equivalent to a transformation characteristic of the *gravitational* force.

It is possible to conceive a supersymmetry theory that is also therefore a theory of gravity. This is actually quite remarkable. The standard model itself does not accommodate gravity at all, and attempts to create a quantum field theory of gravity have led to little more than a hundred years of frustration. By extending the standard model to include supersymmetry, it seems that we open the door to gravity.

Such theories are collectively called *supergravity*. They introduce the gravitino, the superpartner of the graviton, the notional force carrier of quantum gravitation. It seems that supersymmetry not only offers the promise of illuminating the path to a grand unified theory, but may also be a key ingredient in any theory purporting to be a theory of everything.

One of the first theories of supergravity was developed in 1976 by American Daniel Freedman, Dutch physicist Peter van Nieuwenhuizen and Italian Sergio Ferrara. They discovered that some of the problems associated with renormalizing a quantum field theory of gravity were somewhat relieved if supersymmetry was assumed. It seemed that the contributions from terms in the equations that had mushroomed to infinity could be partly offset by terms derived from the gravitino. There was indeed some cancellation, but the problem didn't go away completely.

---

* Actually, a supersymmetry transformation is equivalent to the square root of an infinitesimal translation in space.

For a relatively short time, excitement built up around a version of supergravity based on eight different kinds of supersymmetry. Such theories include not only the particles of the standard model and their superpartners, but many others as well. But these theories could not be renormalized, and by 1984, interest in them was waning.

The reality check

Obviously, SUSY has a lot going for it, and many contemporary theoretical physicists are convinced that nature must be supersymmetric. Of course, the theory will stand or fall on whether or not sparticles are observed in high-energy particle collisions, or supersymmetric WIMPs are detected.

The big problem with SUSY is that the supersymmetry must be broken, and it is not at all obvious how this is supposed to happen. Using a Higgs-like mechanism which ties symmetry-breaking to a real scalar field and a real scalar boson gives incorrect particle masses. Mechanisms for 'soft' supersymmetry-breaking have been devised, but these tend to introduce yet more fields and yet more particles.

This is a bit of a nightmare. As Columbia University mathematical physicist Peter Woit puts it:

> Since one doesn't understand the supersymmetry breaking, to define the MSSM one must include not only an unobserved super-partner for each known particle, but also all possible terms that could arise from any kind of supersymmetry breaking. The end result is that the MSSM has at least 105 extra undetermined parameters that were not in the standard model. Instead of helping to understand some of the eighteen experimentally known but undetermined parameters of the standard model, the use of supersymmetry has added in 105 more.[8]

With so many parameters undetermined by the theory, it becomes impossible to use it to make any predictions. So, there are no real predictions for the masses of the sparticles, for example, other than that some must lie in the range from a few hundred up to a thousand GeV in order to provide a natural fix for the hierarchy problem. As we have seen, there is much riding on the role of the LSP as a candidate dark matter particle, but no SUSY theorist can tell you the identity of the LSP.

I'm afraid there's more. Whenever we establish symmetry relationships between particles, there is always the risk that we get more than we bargained for. Specifically, identities that appear from experiment to be conserved (and, whether by default or not, are conserved in the standard model) become transient. The seemingly impossible becomes possible. And this is the case in SUSY.

As far as we can tell, a muon cannot decay into an electron and a photon. This transformation is not forbidden on energy grounds, but it simply does not happen – it is a process that has never been observed in an accelerator or particle collider. In the context of the standard model, this kind of fact is rationalized in terms of the conservation of muon and electron *number*. We assign a muon (or electron) a muon (electron) number of +1. The anti-muon (positron) is assigned a muon (electron) number of -1. In particle collisions involving muons and electrons we find that muon or electron number is conserved. The total numbers of muons and electrons coming out of such a collision are the same as the numbers of muons and electrons going in.

This situation is maintained in SUSY, until we break the supersymmetry. The large masses of smuons and selectrons cause their identities to blur somewhat, and interactions between muons and electrons and smuons and selectrons can provide a convoluted path in which a muon transforms into an electron. In other words, in a broken SUSY theory, transformations become possible which are not observed experimentally.

There are other problems which need not detain us. It is sufficient at this stage for us to note that SUSY has some uncomfortable consequences. For every standard model problem that it resolves, another problem arises that needs a fix.

Of course, the promise of SUSY is that it provides an important stepping stone on the path towards grand unification. And indeed, a supersymmetric SU(5) theory of the subatomic forces predicts rates of proton decay much more in line with observation. But yet again we get more than we bargained for. A supersymmetric GUT requires a pair of Higgs particles associated with the SU(2) component of the broken theory. It also requires a triplet of colour Higgs particles associated with SU(3). Now, in order to be consistent with our experience of the forces involved, the masses of the Higgs doublet must come out relatively small (about 100 GeV), consistent with the electro-weak energy scale. But the Higgs triplet must have large masses, consistent with the energy scale of grand unification.

There is no obvious way to fix the mass difference of the Higgs doublet and Higgs triplet naturally from within the theory. But if it isn't fixed, the Higgs triplet can mediate proton decay and we're back at square one: with protons decaying faster than we observe. This is generally known as the doublet–triplet splitting problem.

It is the hierarchy problem all over again.

Weighing the evidence

So, where do we stand? There can be little doubting the value of SUSY in terms of the logic of the approach and its promised resolution of some of the fundamental problems of the standard model. Our reality check may cause us some discomfort, but nobody ever said this was going to be easy.

Is nature supersymmetric? Of course, our seemingly protracted debates are immediately ended the moment we find unambiguous evidence for sparticles. Gordon Kane again:

> If the superpartners are found, it will confirm that supersymmetry is part of our description of nature. If superpartners are not observed, it will show that nature is not supersymmetric.[9]

Now, it would be asking too much of a theory with 105 additional parameters to come up with hard-and-fast predictions for the masses of the sparticles, but it's enough for us to know that at least some of them should have masses of the order of hundreds of GeV. The theory stands or falls on its prediction of sparticles, at whatever masses they can be shown to possess. For this reason, although generally sceptical, I tend to regard SUSY as a legitimate theory of physics. It is at least testable, in the sense of the Testability Principle, although it does have a fairy-tale tendency, as we will see.

Kane's book on supersymmetry from which I have quoted was published in 2000, and at this time there were plenty of reasons for optimism. The lighter sparticles were thought to be in range of CERN's Large Electron-Positron (LEP) Collider and Fermilab's Tevatron. If these colliders came up empty, there was the promise of the LHC, CERN's successor to the LEP, which would be designed to achieve total proton–proton collision energies of 14 TeV.

This sounds perfect, but don't be misled. Not all of the headline collision energy can be utilized. Protons consist of quarks and gluons, and the energy of a 7 TeV proton travelling very near the speed of light around a collider is distributed over these components. Proton–proton collisions are actually quark–quark, gluon–gluon or quark–gluon collisions, and the energies of these can be a lot less than the headline collision energy.

Nevertheless, when the LEP and the Tevatron did indeed come up empty, the LHC became the gaming house in which the SUSY gamble would either pay out, or not. American theorist Lisa Randall put it this way:

> … if supersymmetry solves the hierarchy problem, it will be an experimental windfall. A particle accelerator that explores energies of about a TeV (1,000 GeV) will find, in addition to the Higgs particle, a host of supersymmetric partners of standard model particles. We should see gluinos and squarks, as well as sleptons, winos … a zino and a photino … With sufficient energy, these particles would be hard to miss. If supersymmetry is right, we will soon see it confirmed.[10]

It's a simple fact that in order for the stop (meaning the stop squark) to stabilize the Higgs mass and provide a natural resolution of the hierarchy problem, it would need to possess a mass not very different from the electro-weak energy scale – a few hundred GeV. This is well within the range of the LHC, even though, at the time of writing (July 2012), the collider had not yet reached its design collision energy of 14 TeV.

If they exist, there are several ways in which stop squarks might be produced in proton–proton collisions in the LHC. They can in theory be produced directly from proton–proton (actually, gluon–gluon) collisions, but the rate of production by this route is thought to be rather limited. A more productive route is possible indirectly via gluinos. A proton–proton collision produces two gluinos, each of which goes on to decay into a top–stop pair.

According to the MSSM, the stop is expected ultimately to decay into a top quark plus the LSP. So, if the collision produces two gluinos, these decay first into two top quarks and two stop squarks and the two stops decay further to give two more top quarks and two LSPs. The end result is a collision producing four top quarks and lots of 'missing' energy, as the LSP behaves like a neutrino and escapes undetected. There are few standard model processes that can give rise to four top quarks, so this gluino-mediated route is an attractive candidate in the search for the stop, as there should be little contribution from 'background' processes.

Alternatively, the stop could decay into a W particle and a sbottom squark, with the sbottom going on to decay into a bottom quark plus the LSP. Irrespective of the actual decay paths, the end result is a number of top and/or bottom quarks, and lots of missing energy.

From the beginnings of proton physics at the LHC in March 2010, the principal focus of attention for the two general-purpose detector collaborations, ATLAS and CMS, has been the search for the Higgs boson. Nevertheless, both collaborations have periodically reported on the search for sparticles and other kinds of 'new physics'. In all cases, no significant excess of events over and above the expected background from standard model processes has been observed.

Absence of evidence cannot necessarily be taken in this case as evidence of absence. At the time of writing, both ATLAS and CMS have yet to analyse all the data gathered from some 350 trillion 7 TeV proton–proton collisions in 2011 and substantially more data at 8 TeV gathered in 2012. However, the signs are not good. The data that have been analysed are not producing the telltale signatures of stop or sbottom squarks, and exclusion limits are being pushed to higher and higher mass ranges. The exclusion limits are starting to become incompatible with a natural resolution of the hierarchy problem.

It is looking increasingly likely that the gamble will be lost. And soon.

The search for WIMPs

Where does this leave us on the question of dark matter? Of course, if nature turns out not to be supersymmetric, then there is no LSP and no SUSY dark matter candidate. This does not mean that there are no WIMPs, however. If we could ever identify a WIMP, its properties might give us important clues to the nature of physics beyond the standard model.

Physicists need relatively little motivation to start a search if the technology (and, more importantly, the funding) is available to support it. The Cryogenic Dark Matter Search (CDMS) experiment is located half a mile underground at the Soudan mine in northern Minnesota. It consists of thirty detectors cooled to temperatures very near absolute zero. Signals generated by a particle interacting inside one of the detectors allow the experimenters to tell whether this is a conventional standard model particle or a WIMP.

On 17 December 2009, the CDMS collaboration announced that it had found just two decay events that were potential candidates involving WIMPs. Two events are insufficient to discriminate positive results from false positives, and further results released in 2011 showed no evidence for WIMPs with a mass below 9 GeV.

A search by the XENON100 experiment at the Gran Sasso National Laboratory in Italy reported similar results in April 2011. This experiment looks for telltale signals from WIMPs interacting with liquid xenon. Data gathered from 100 days of operation between January and June 2010 turned up three candidate events. This would sound promising were it not for the fact that the number of background events (false positives) in this period was predicted to be nearly two.

On 18 July 2012, the XENON100 experiment reported the results of a further 225 days of data-taking during 2011 and 2012, with even higher instrument sensitivity. Once again, the two observed candidate WIMP events could not be distinguished from the background (predicted to be one event). Instead, the experiment was able to extend the mass range for which WIMPs can be confidently *excluded*, pushing the hypothetical particles up to higher and higher masses.

It seems we cannot look to WIMPs to provide us with guidance just yet.

The end of SUSY?

It is a mistake to think that scientific progress is driven solely on the basis of scientific evidence. Theorists who have become enamoured of a particular theoretical structure may argue in its favour long after unambiguous evidence against it has been published and broadly accepted. In his famous dissection of the structure of scientific revolutions, Thomas Kuhn drew analogies with the process and aftermath of political revolution. Long after a revolution has sealed the fate of a failed political system, it will always be possible to find dissidents who will argue that things were better in the 'old days'.

It is this attitude among some theorists that is beginning to edge SUSY in the direction of fairy-tale physics. Instead of accepting the evidence at face value and acknowledging that nature might not, after all, be supersymmetric, they work to develop alternative interpretations or extensions to the theory which explain why the absence of light sparticles and WIMPs is not inconsistent.

For example, *split supersymmetry*, first proposed in 2003, pushes the sparticle masses to higher energies, above a TeV, and so out of reach of our ability to detect them at the LHC. In this form, SUSY is spared the embarrassment of failing the current experimental tests by making the tests no longer relevant. But in relaxing the requirements of 'naturalness', such theories no longer provide a solution to the hierarchy problem.

There is obviously a limit to the number of times that theorists can play this game. But such argumentation doesn't necessarily end even when the evidence is overwhelming. I've been struck by the enduring popularity of Fritjof Capra's book *The Tao of Physics*, which still makes an occasional appearance on the shelves of my local bookstore 37 years after it was first published.

The Tao of Physics extols the virtues of the 'bootstrap' theory, which was very popular in the 1960s. This theory argued in favour of a kind of 'nuclear democracy', in which there are no 'elementary' particles as such. Instead, all of the particles known to physicists are supposedly formed from combinations of each other. This model quickly went out of fashion when evidence for quarks – the 'elementary' constituents of protons, neutrons and mesons – began to emerge in the late 1960s.

Despite the fact that the theory has long since ceased to be regarded as having anything to do with the current version of reality, it appears there is still a readership for a popular book about it. No doubt this has something to do with the parallels that Capra draws between the bootstrap theory and aspects of Eastern religious philosophies.

Is this the ultimate fate of SUSY? It is perhaps a little too early to tell, but further proton–proton collision data collected at the LHC in 2012 should in principle prove decisive. My belief is that many theorists will continue to declare that the writing of obituaries for SUSY is premature, finding ever more convoluted and elaborate ways of keeping the theory alive. Like Darth Vader's journey to the dark side, SUSY's journey to fairy-tale physics will then be complete.

There is a simple reason for this: there is now far too much at stake. The post–standard model structure that has dominated theoretical particle physics for more than thirty years is, of course, superstring theory.

And what, precisely, is it that makes superstrings 'super'?

SUSY, of course.

In the Cemetery of Disappointed Hopes

Superstrings, M-theory and the Search for the Theory of Everything

I'm still working, passionately, though most of my intellectual off-spring are ending up prematurely in the cemetery of disappointed hopes.

Albert Einstein[1]

Soon after his triumphant presentation of the general theory of relativity to the Prussian Academy of Sciences in 1915, Einstein began searching for a way to unify gravity and electromagnetism. At this stage the quantum was much more than a hypothesis but it was not yet a fully fledged theory. The development of quantum theory still awaited the intellectual energies of Bohr, Heisenberg, Pauli, Schrödinger, Dirac and Born, and many others.

And yet Einstein was already acknowledging that such a unification would require that the general theory of relativity should somehow be 'reconciled' with quantum theory.

In truth, as the discomforting consequences of quantum theory began to emerge, bringing problems such as quantum probability, the collapse of the wavefunction and 'spooky' action-at-a-distance, Einstein came to believe that a proper unified theory would actually serve to eliminate these problems, a belief that would shape the nature of his intellectual pursuit over the later stages of his life. This despite the fact that experimental science of the 1930s and 1940s established beyond any real doubt the supremacy of the quantum and the importance of two further forces – the strong and weak nuclear forces – which would need to be accommodated in any theory purporting to be a unified theory of everything.

Einstein's pursuit of a unified theory of gravity and electromagnetism was to be largely fruitless. His sense of growing frustration is captured eloquently in the title quote, from a letter to his colleague Heinrich Zangger which he wrote in 1938. And despite the immense progress in theoretical physics since Einstein's death in 1955, the search for a theory of everything continues to carry a strong, unshakeable aura of futility.

It is hard to resist the conclusion that the search for the theory of everything remains firmly locked in Einstein's cemetery of disappointed hopes. It seems that only a passion for fantasy and fairy tales is keeping the search alive, at least in one specific direction.

Cylinder world: physics in five dimensions

Einstein's spirits were temporarily lifted by one idea from a young German mathematician called Theodor Kaluza. In 1919, Kaluza showed how it might be possible to combine gravity and electromagnetism by extending the four-dimensional framework of spacetime to include an additional, fifth dimension.

As a mathematician, Kaluza was used to dealing with abstract 'dimensions' and was not overly concerned about how this fifth dimension should be interpreted physically. He simply found that solving Einstein's field equations in five dimensions instead of the usual four resulted in the natural and spontaneous emergence of Maxwell's equations for electromagnetism.

However, he did require the extra spatial dimension to be circular, such that a particle following a straight trajectory along this dimension would soon find itself back where it started.* What appears as a straight line in our conventional, Euclidean geometry of up–down, left–right, back–front becomes a circle in Kaluza's additional dimension. Einstein wrote to him in April 1919: 'The idea of achieving [a unified theory] by means of a five-dimensional cylinder world never dawned on me ... At first glance I like your idea enormously.'[2]

---

* If we recall that QED, the quantum field theory of electromagnetism, is based on the circular $U(1)$ symmetry group, then we might get some sense of the strong connection between 'circularity' and electromagnetism.

Einstein helped Kaluza to get his work published, although for reasons that are not clear, this took a further two years. Kaluza's ideas appeared in print in 1921.

Oskar Klein was by contrast a physicist. He had trained as a physical chemist under the great Swedish Nobel laureate Svante Arrhenius in Stockholm, before moving to Copenhagen in 1917 to work with Niels Bohr. He returned to his native Sweden to complete his doctorate in 1921. Five years later he arrived independently at Kaluza's idea but tried to give it at least some kind of physically realistic, and specifically quantum, interpretation. He also noted that if this dimension were to be rolled up small and tight, with the radius of the cylinder measurable only on sub-nuclear scales, then perhaps this was the reason we could not perceive it directly.

Such a 'hidden' dimension forms what is known as a *compact set*. The process of transforming a dimension so that it curls up tightly in a closed structure or 'manifold' instead of stretching off to infinity is called *compactification*.

Klein's reinterpretation and modification of Kaluza's idea became known as Kaluza–Klein theory. Einstein was impressed, but was now also expressing some misgivings about the idea. 'Klein's paper is beautiful and impressive,' he told his colleague, Austrian physicist Paul Ehrenfest, 'but I find Kaluza's principle too unnatural.'[3] It seems he regarded it as somewhat curious to introduce a new dimension only to tie it up artificially and hide it away in order to account for the simple fact that we don't experience it.

Einstein went on to develop alternative theoretical structures that did not rely on compactification, but these were broadly unsuccessful.

For a time, Klein was encouraged by the thought that the extra spatial dimension might lie at the heart of all quantum phenomena. But the idea was largely forgotten as quantum theory and the Copenhagen interpretation became firmly established.

Of strings and superstrings

And so we turn to theories of strings and superstrings.

In this instance, I think that a broadly chronological tale will help us to understand the basis on which superstring theory has been constructed. Specifically, it will allow us to appreciate that, in seeking to resolve problems associated with its applicability, the theorists have been obliged to pile assumption on top of assumption.

We will then be in a position properly to judge the theory for ourselves.

So, fast-forward some 42 years, from Klein's paper on a five-dimensional unification of gravity and electromagnetism to Gabriele Veneziano, a young Italian postdoctoral physicist working at CERN in the summer of 1968. Whilst puzzling over ways to describe the scattering of mesons (such as pions) using a theory of strong force interactions, Veneziano developed a mathematical relationship to calculate certain scattering 'amplitudes'. These amplitudes represent the probabilities for the production ('scattering') of two final mesons from two initial mesons at different collision angles. They became known as Veneziano amplitudes.

There was no strong theoretical basis for the formula that Veneziano had devised, but it was quite familiar. It was Swiss mathematician Leonhard Euler's 'beta function', or Euler's 'integral of the first kind'.

When young Yeshiva University professor Leonard Susskind heard about this relationship, he was struck by its simplicity and decided to try to understand where it had come from. As he later explained:

> I worked on it for a long time, fiddled around with it, and began to realize that it was describing what happens when two little loops of string come together, join, oscillate a little bit, and then go flying off. That's a physics problem that you can solve. You can solve exactly for the probabilities for different things to happen, and they exactly match what Veneziano had written down. This was incredibly exciting.[4]

At this time, the 'strings' in question were identified with quarks tethered to anti-quarks* by the 'elastic' of the strong force. Similar discoveries were made independently by Danish physicist Holger Nielsen at the Niels Bohr Institute in Copenhagen and by Yoichiro Nambu in Chicago.

---

* Remember that mesons are formed from quark–anti-quark pairs.

One thing led to another. What if, instead of treating elementary particles as point particles, we instead treat the elastic that holds them together as 'fundamental'? What if the string's the thing?

In such a string theory the different particles would not be translated into different kinds of strings, but into different *vibrational patterns* in one common string type. The mass of a particle would then be just the energy of its string vibration, with other properties such as charge and spin being more subtle manifestations.

It was a breathtaking idea. Switching from abstract point particles to extended one-dimensional strings offered the promise that it might be possible to avoid all the problems with infinities and renormalization that had plagued quantum field theory from its inception. But this was an *idea* nonetheless.

Perhaps it would help to keep track of these, as we accumulate them.

The String Assumption. *The original string theory is founded on the assumption that elementary particles can be represented by vibrations in fundamental, one-dimensional filaments of energy.*

Now, to be completely fair, we should acknowledge that quantum field theory, and hence the entire structure of the standard model of particle physics, is similarly founded on an (often unstated) assumption: that of point particles. The difference is that, within its recognized limits, we know that the standard model *works*, and this helps to justify the point particle assumption even though we know that in principle it can't be right, or at least it can't be the whole story. To justify changing the basis of this assumption to one of strings, we would need some confidence that this leads to a string theory that is demonstrably *better* than the standard model; a string theory that solves at least some of the standard model's problems.

But early versions of string theory did not look at all promising, and consequently did not attract much attention. The theory was derived from mathematical relationships established for the scattering of 'scalar' particles, such as mesons, with zero spin. It could deal only with strings that describe bosons. The theory also required as many as 26 spacetime dimensions, 25 dimensions of space and one of time.

Worst of all, the theory also predicted the existence of tachyons, hypothetical particles that possess imaginary mass and travel only at speeds faster than light. Tachyons wreak havoc on the principles of cause and effect and are anathema to any theoretical structure with pretensions to describe the real world. The presence of tachyons is a sign that something, somewhere, has gone horribly wrong.

Most theorists turned their attentions away from string theory. But a few persisted, enamoured of its potential and reluctant to throw the baby out with the bathwater. By 1972, one of the big problems of early string theory had been solved, by French theorist Pierre Ramond, working at the National Accelerator Laboratory in Chicago, and by American John Schwarz and French theorist André Neveu at Princeton University.

You can probably figure out for yourself roughly how this was done. How do you get fermions into a theory that describes only bosons? By assuming a fundamental symmetry between fermions and bosons, such that for every boson there is a corresponding fermion. In other words, by assuming that the string universe is supersymmetric.

Now this is a greatly oversimplified way of looking at the development of superstring theory; as is typically the case, the true history is not quite so neat. Ramond had written a paper concerned with non-interacting fermionic strings; Schwarz and Neveu were working on the theory of bosonic strings when they realized that these could be brought together. As Schwarz later explained:

> So we got ahold of [Ramond's] paper just about the time we were finishing ours, and we were struck by the mathematical similarity of what we had done and what he had been doing. Then we realized that Ramond's fermionic strings could be made to interact with our bosonic strings and it was all part of a consistent theory. Over the years, we've used different names for this theory, and it has evolved into what we today call superstrings.[5]

SUSY was emerging at around the same time, and some have argued that it was actually derived from early string theory. It doesn't matter much precisely how it all came about, but we do need to acknowledge another important layer of assumption. Although we encountered the supersymmetry assumption in the last chapter, it is important to repeat it here in the context of string theory.

The Supersymmetry Assumption. *Superstring theory is based on the assumption that there exists a fundamental spacetime symmetry between matter particles (fermions) and the force particles (bosons) that transmit forces between them.*

Schwarz found that the number of spacetime dimensions required for superstrings to vibrate in was no longer the 26 demanded by string theory, it was just ten: nine dimensions of space and one of time. Although this was still a lot more than the four dimensions of our everyday experience, it was at least a step in the right direction.

And – good news – the tachyons were now gone.

Superstrings and the graviton

Despite these advances, there was no great rush to embrace superstring theory. In 1971, Gerard 't Hooft and Martinus Veltman showed that quantum field theories incorporating the Higgs mechanism could in general be renormalized. Two years later, the notion of asymptotic freedom was demonstrated to be compatible with quantum field theory, and this gave birth to what was eventually to become QCD. Within a few years, the structure of the standard model of particle physics was established.

There seemed little to be gained from further work on superstrings as a theory of the strong nuclear force. The few hundred or so theorists who had worked on early string theory moved on to other things.

But Schwarz was still not ready to let go. At Princeton he had collaborated with two French theorists, André Neveu and Joël Scherk. Schwarz's work on superstrings had not earned him a tenured professorship at Princeton University, in itself a sign that the theory was not highly regarded. He accepted the offer of a research associate position at the California Institute of Technology (Caltech) in Pasadena. Scherk visited Caltech in January 1974 and they continued their collaboration. Schwarz explained why:

> I think we were kind of struck by the mathematical beauty [of superstring theory]; we found the thing a very compelling structure.

I don't know that we said it explicitly, but we must have both felt that it had to be good for something, since it was just such a beautiful, tight structure. So, one of the problems that we had had with the string theory was that in the spectrum of particles that it gave, there was one that had no mass and two units of spin. And this was just one of the things that was wrong for describing strong nuclear forces, because there isn't a particle like that. However, these are exactly the properties one should expect for the quantum of gravity.[6]

If the challenge to provide a theoretical description of the strong nuclear force had been met, then the challenge to provide a quantum theory of gravity had not. It seemed that superstring theory not only promised to accommodate all the elementary particles of the standard model then known; it also predicted a particle with the properties of the graviton.

Superstring theory was not a theory of the strong force. It seemed that it was potentially a theory of *everything*.

Superstrings come in two forms: open and closed. Open strings have loose ends that we can think of as representing charged particles and their anti-particles, one at either end, with the string vibration representing the particle carrying the force between them. Open strings therefore predict both matter particles *and* the forces between them. However, the theory also *demands* closed strings. When a particle and anti-particle annihilate, the two ends of the string join up and form a closed string.

But if there are closed strings, there are also gravitons and the force of gravity. Rather than trying to shoehorn quantum theory into general relativity, or vice versa, superstring theory appeared to be saying that all nature's forces are just different vibrational patterns in open and closed strings. In superstring theory these forces are *automatically* unified.

Superstring theory appeared to offer great promise. It seemed that all the elementary particles then known – their masses, charges, spins, the forces between them and all the standard model parameters that could not be derived from first principles – could be subsumed into a single theory with just two fundamental constants. These were the constants that determine the tension of the string and the coupling between strings.

Nobody was interested.

Enter Witten: the first superstring revolution

Schwarz continued his collaboration with both Neveu and Scherk. They worked on a variety of aspects of superstring theory and explored the possibility that the extra spatial dimensions demanded by the theory might somehow be responsible for spontaneous supersymmetry-breaking.

Supersymmetry was also becoming established as a four-dimensional extension of standard model quantum field theories. Supergravity was evolved, and Schwarz contributed to its development.

Whilst working for a few months at CERN in Geneva, Schwarz began collaborating with British physicist Michael Green, based at Queen Mary College in London. Together they explored aspects of three different kinds of superstring theory. These became known as Type I, Type IIA and Type IIB. All require ten spacetime dimensions but differ in the way that supersymmetries are applied. Like the MSSM, Type I superstring theory makes use of one supersymmetry. The Type II theories use two.

Although their work continued to be largely ignored, the theory was beginning to win a few advocates. Among them was Princeton mathematical physicist Edward Witten.

By the early 1980s, Witten was still relatively young but was already a force to be reckoned with. He had had a rather eclectic career. After studying history and linguistics at Brandeis University near Boston, he went on to read economics at the University of Wisconsin and embarked on a career in politics, working on George McGovern's 1972 presidential campaign.

After McGovern's overwhelming defeat by Richard Nixon, he abandoned politics and moved to Princeton to study mathematics. He migrated to physics shortly afterwards, following in the footsteps of his father, Louis, professor of physics at the University of Cincinnati. He studied for his doctorate under David Gross, securing his PhD in 1976. Just four years later he was a tenured professor at Princeton.

Witten was establishing a reputation as a bona fide genius, a modern-day Einstein. In 1982 he was awarded a MacArthur Foundation 'genius' grant. Young Princeton graduate Peter Woit described the impression of being in contact with such a vastly superior intelligence. Following some paces behind Witten as he crossed the Princeton campus, Woit climbed the steps to the large plaza in front of the library:

When I reached the plaza [Witten] was nowhere to be seen, and it is quite a bit more than thirty feet to the nearest building entrance. While presumably he was just moving a lot faster than me, it crossed my mind at the time that a consistent explanation for everything was that Witten was an extra-terrestrial being from a superior race who, since he thought no one was watching, had teleported back to his office.[7]

Witten's involvement in superstring theory was in itself sufficient to draw attention to the subject. He quickly persuaded Schwarz and Green that if superstring theory was to become a viable alternative to the standard model and a serious candidate as a theory of everything, then they needed to demonstrate that it was free of *anomalies*.

An anomaly occurs when the symmetry of the theory breaks down as a result of making so-called 'one-loop' quantum corrections. If this happens, the mathematical consistency of the theory is lost and there's a chance that it might no longer make consistent predictions (for example, such an anomaly in QED might lead to the prediction of photons with three spin orientations or three polarization directions). Needless to say, in the quantum field theories of the standard model, all such anomalies cancel.

Different versions of superstring theory have different symmetry properties. Type IIA is mirror-symmetric, and the physicists could be confident that this theory would exhibit no residual anomalies. But look in a mirror. The world we inhabit is not mirror-symmetric.

In 1984, Witten and Spanish physicist Luis Alvarez-Gaumé pointed out that further anomalies would arise in superstring theory due to the gravitational field. But they went on to show that in a low-energy-approximation Type IIB theory, these anomalies did indeed cancel. This was encouraging, but still left a question mark over the Type I theory.

In the summer of 1984, Green and Schwarz discovered that in a low-energy-approximation Type I superstring theory based on the symmetry group SO(32), a group of rotations in 32 dimensions,* the anomalies did indeed all cancel out. Witten picked up rumours of their

---

* Not to be confused with the 10 space-time dimensions of the theory.

breakthrough, and called to find out more. They sent him a draft manuscript by Federal Express.

Things then happened very quickly.

Green and Schwarz published their paper in September 1984. Witten submitted his first paper on superstrings to the same journal later that same month.

At Princeton, David Gross, Jeffrey Harvey, Emil Martinec and Ryan Rohm (who would collectively come to be known as the Princeton String Quartet) found yet another version of the theory, called heterotic (or hybrid)[*] superstring theory, in which the anomalies also cancelled. It turned out that there were two types of heterotic superstring theory, one based on the symmetry group formed from the product $E_8 \times E_8$, where E_8 is an algebraic group with 248 dimensions, and the other based on the symmetry group SO(32). Both theories require ten spacetime dimensions. They submitted their paper for publication in November 1984.

The first superstring revolution had begun.

Hiding the extra dimensions: Calabi–Yau spaces

Kaluza discovered that by solving Einstein's field equations in a five-dimensional spacetime, Maxwell's equations would emerge naturally. Klein deduced that by rolling up – compactifying – the extra spatial dimension into a cylinder with a radius of sub-nuclear scale, its role in the theory could be preserved without the embarrassment of having to experience it directly.

Superstring theory couldn't settle for just one extra spatial dimension. It demanded an extra six. The reason is not so difficult to understand, once you have accepted the assumption that unification can be achieved by expanding the number of dimensions. Kaluza–Klein theory uses five dimensions to join gravity and electromagnetism, to which we must now add further dimensions to accommodate the weak and strong nuclear forces. Alternatively, if we make the string assumption, then we need enough 'degrees of freedom' for the strings to vibrate in

---

[*] Heterotic superstring theories are hybrids of Type I and bosonic superstring theories.

if these vibrational patterns are to represent all the elementary particles and all the forces between them.

The point about superstring theory is that it demands precisely nine spatial dimensions, no more and no fewer. But what must we now do with this embarrassment of riches? A lifetime of experience tells me that the world is stubbornly three-dimensional and I have little doubt that I will not find another six dimensions no matter how hard I look.

I feel another assumption coming on.

The Compactification Assumption. *The nine spatial dimensions demanded by superstring theory can be reconciled with our experience of a three-dimensional world by assuming that six dimensions are compactified into a manifold with a size of the order of the Planck length, about 1.6 billionths of a trillionth of a trillionth (1.6×10^{-33}) centimetres.*

Klein had rolled up the extra dimension into a cylinder, but a further six dimensions demand a much more complex structure. In 1984, American theorist Andrew Strominger, then at the Institute for Advanced Study in Princeton, searched for an appropriate structure for this manifold in collaboration with British mathematical physicist Philip Candelas, then at the University of Texas. This was not a free choice. Supersymmetry places some fairly rigid constraints on precisely how these extra dimensions can be rolled up. Strominger's search led to the library, and a recent paper by Chinese-born American mathematician Shing-Tung Yau. The paper contained a proof of something called the Calabi conjecture, named for Italian-American mathematician Eugenio Calabi.

As Strominger explained: 'I found Yau's paper in the library and couldn't make much sense of it, but from the little I did understand, I realized that these manifolds were just what the doctor ordered.'[8]

Calabi's conjecture concerns the geometric structures that are allowed by different topologies, the various shapes that 'mathematical' spaces can possess. Although this was a problem in abstract geometry, solutions to such problems have been of interest to physicists ever since Einstein established the connection between geometry and gravity in the general theory of relativity. One interpretation of the Calabi conjecture is that it suggests that in certain spaces, gravity is possible even in the absence of matter.

Yau developed a proof of the conjecture, which he discussed with Calabi and Canadian-born American mathematician Louis Nirenberg on Christmas Day 1976.[*] The proof confirmed the existence of a series of shapes – now called Calabi–Yau spaces – that satisfy Einstein's field equations in the absence of matter. It was indeed just what the doctor ordered.

Strominger and Candelas got in touch with Gary Horowitz, at the University of California in Santa Barbara, a physicist who had worked with Yau as a postdoctoral associate. Strominger also visited Witten, to discover that the latter had independently arrived at the same conclusion. At Strominger and Witten's request, on a flight from San Diego to Chicago Yau worked out the structure of a Calabi–Yau space that would generate three families or generations of matter particles.

The four theorists collaborated on a paper which was published in 1985. Thus was born the idea of 'hidden dimensions'. If I mark an infinitesimally small point on the desk in front of my keyboard, and could somehow zoom in on this point and magnify it so that a distance of a billionth of a trillionth of a trillionth of a centimetre becomes visible, then superstring theory says that I should perceive six further spatial dimensions, curled up into a Calabi–Yau shape (see Figure 8).

On the one hand, this is a perfect example of how we make progress in science. We know that abstract point particles lead to problems. Abstract one-dimensional strings appear to offer better prospects. We reach for supersymmetry because we want a theory that describes fermions as well as bosons, and this eliminates some of the problems of the original string theory (such as tachyons) and yields theories free of anomalies. Superstrings demand a ten-dimensional spacetime, so we borrow concepts from mathematics and tuck the six extra spatial dimensions out of sight in a Calabi–Yau space. We can find a Calabi–Yau space that is consistent with three generations of elementary matter particles. This all seems perfectly logical and reasonable.

---

[*] Apparently this was the only time that all three were available.

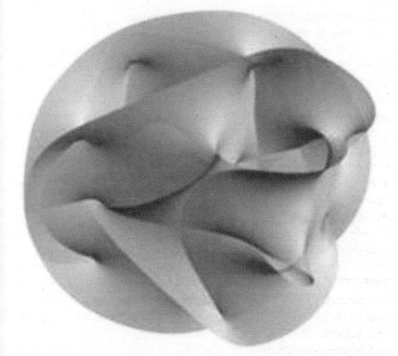

Figure 8 The Calabi–Yau manifolds or Calabi–Yau shapes are complex, high-dimensional algebraic surfaces. They appear in superstring theory as manifolds containing the six additional spatial dimensions required for the strings to vibrate in. Because they have dimensions of the Planck length, these manifolds are far too small to be visible. Source: Wikimedia Commons.

But then there is the other hand. On what basis do we choose strings, as opposed to any other kind of abstract construction? Because of a (possibly rather tenuous) relationship between the behaviour of strings and the beta function identified in the scattering of mesons by Veneziano. On what basis do we assume supersymmetry between fermions and bosons? Because this is the only way to get both kinds of particles into the same picture. What is the basis for assuming that six dimensions must be hidden in a Calabi–Yau space? Because it is our experience that the universe is four-dimensional.

I think you get the point. Although this is all perfectly logical and reasonable, what we are actually doing is piling one grand assumption on top of another. I want to emphasize that there is no experimental or observational basis for these assumptions. This is a theory with little or no foundation in empirical reality. It is rather a loose assemblage of assumptions, ideas or hypotheses that say rather more about how we would like the universe to be than how it really is.*

I don't think it's uncharitable to suggest that this is looking increasingly like a house of cards.

Of course, the Theory Principle tells us that this is still okay. Science is a very forgiving discipline in that it never really matters overmuch *how* you arrive at a theory. If it can be shown to work better than existing alternatives, then no matter how much speculation or luck was involved, you can sit back and wait patiently for the Nobel Prize committee to reach the right decision.

Before we take a look at superstring theory's predictions, we have to deal with the fact that there appear to be at least five different versions of the theory – Type I, Type IIA, Type IIB and two versions of heterotic superstring theory. This is a little embarrassing for a theory that has pretensions to be *the* theory of everything.

Resolving this issue would require the biggest assumption of all, one that sparked the second superstring revolution.

The M-theory conjecture and the second superstring revolution

As theories proliferated and superstring theory lost any sense of uniqueness, interest began to wane. British superstring theorist Michael Duff explained it like this:

> Theorists love *uniqueness*; they like to think that the ultimate *Theory of Everything* will one day be singled out, not merely because all rival theories are in disagreement with experiment, but because they are mathematically inconsistent. In other words, that the universe is the way it is because it is the only possible universe.[9]

---

* Perhaps it is more correct to say that the theorists would prefer a universe like this because it is one that they can develop a theory to describe.

There were also rumblings from some theorists that, based on calculations using supergravity, the right number of spacetime dimensions is actually eleven, not ten. It all seemed to be getting rather out of hand. Then, at a superstring theory conference at the University of Southern California in March 1995, Witten made a bold conjecture. Perhaps the five different ten-dimensional superstring theories are actually instances or approximations of a single, overarching eleven-dimensional structure. He called it M-theory. He was not specific on the meaning or significance of 'M'.

This was a conjecture, not a theory. Witten demonstrated the equivalence of a ten-dimensional superstring theory and eleven-dimensional supergravity but he could not formulate M-theory; he could only speculate that it must exist.

The M-Theory Conjecture. *We assume that the five variants of superstring theory can be subsumed into a single, eleven-dimensional framework. But nobody has yet been able to write this theory down on paper.*

This is a simple fact that many readers of popular presentations of superstring theory somehow tend to miss. *M-theory is not a theory.* Nobody knows what M-theory looks like, although many theorists have tinkered with structures that they believe it could or should possess. So, on top of a foundation built from a sequence of assumptions, we now erect the biggest assumption of all. We assume that a unique eleven-dimensional superstring theory is possible in principle, although we don't yet know what this theory is.

I think this is a truly remarkable state of affairs. Of course, this kind of speculative theorizing goes on all the time in physics, and there are plenty of examples from history. We might look again at the sequence of assumptions that have led us here and conclude that theorists who commit themselves to M-theory are either brave or foolhardy or both. We might express some concern for their future career development, but we might be ready and willing to acknowledge that the very idea of academic freedom means that there will always be some few committed to what seem to us to be mad or trivial pursuits.

But M-theory is not the preserve of a few theorists who have become addicted to its beauty and the tightness of its structure, as Danish historian Helge Kragh explains:

> With M-theory and what followed, new recruits were attracted to the field. It is estimated that the string community amounts to some 1,500 scientists worldwide, a remarkably large number given the abstract and purely theoretical nature of string physics. Incidentally, this means that there are as many string theorists today as the total number of academic physicists in the world about 1900, all countries and fields of physics included.[10]

Given that we can trace the first superstring revolution back nearly thirty years and the second revolution began seventeen years ago, we might well ask what all these theorists have been up to during this time.

One of the big revelations of Witten's M-theory conjecture concerns so-called *dualities* that prevail between different versions of superstring theory. For example, S-duality, or strong–weak duality, allows states with a coupling constant of a certain magnitude in one type of theory to be mapped to states with a coupling given by the reciprocal of this constant in its dual theory. This means that strong couplings which reflect the influence of strong forces in one theory would be weak couplings in the dual theory. So, the perturbation theory techniques used so successfully in QED, normally applicable only for systems of low energy and weak coupling, could now be applied to the dual theory to deal with problems involving high-energy, strong-coupling regimes.

A lot of previously intractable calculations suddenly became tractable.

Branes and braneworlds

There was a lot more, however. Introducing the extra dimension in M-theory opened up an extraordinarily rich structure of mathematical objects. Superstring theory was no longer limited to one-dimensional strings. It could now accommodate higher-dimensional objects, called *membranes*, or 'branes', which quickly took over from strings as the primary focus of investigation, leading Duff to refer to it as 'the theory formerly known as strings'.[11]

Branes may have up to nine dimensions. A string is a 'one-brane'. Two-dimensional membranes (sheets) are referred to as two-branes. Three-branes are actually three-dimensional spaces. Generically, these objects are referred to as p-branes, where p refers to the number of dimensions.

Branes lie at the heart of superstring theory's dualities. For example, particles in one version of the theory become branes in the dual theory. This kind of interrelationship lends credibility to the assumption that these are more than just mathematical curiosities – that branes reflect or represent physically real properties and behaviours of our universe. This is a perfectly logical inference, but we should be clear once again that there is no single piece of experimental or observational evidence to suggest that the fundamental elements of reality are multidimensional membranes.

Branes are the principal objects of a theory which is conjectured to exist but which has yet to be written down. Exploring the mathematical relationships between and involving branes in various forms and connecting these with physically meaningful properties and behaviours requires another big assumption.

The Brane Assumption. *The higher-dimensional mathematical objects that arise in M-theory are assumed to have physical significance: they are assumed to describe aspects of empirical reality.*

A subcategory of p-branes, called 'D-branes', are particularly interesting because they represent locations in space where open strings can end.* Such objects not only represent different ways in which the various spatial dimensions can be occupied; they also possess shape and particle-like properties such as charge. Branes are dynamic objects; they move around and interact and are susceptible to forces.

In fact, Witten and Czech theorist Petr Hořava constructed a model universe from two parallel ten-dimensional D-branes separated in the direction of the eleventh dimension. The space between branes is referred

---

* The 'D' derives from Peter Dirichlet, a nineteenth-century German mathematician.

to as the 'bulk'. They demonstrated that this structure is equivalent to a strong-coupling version of the original heterotic $E_8 \times E_8$ superstring theory first devised by the Princeton String Quartet.

The fascination with such 'braneworlds' derives from the fact that models can be created in which all material particles, which are composed of open strings, are permanently fastened to one or other of the D-branes, the spaces where the open strings end. Models can be set up so that these particles cannot detach from the brane and so cannot explore other spatial dimensions that exist 'at right angles' to the dimensions of the D-brane. In other words, the particles cannot travel 'at right angles to reality'.

A ten-dimensional D-brane can be thought to consist of three 'conventional' spatial dimensions which extend off to infinity, six dimensions curled up and tucked away in a Calabi–Yau space, and time. The model demands that, once fixed to the D-brane, all the material particles of the standard model are then constrained to move in the three-dimensional space that we ourselves experience.

But particles formed from closed strings, such as the graviton, are not so constrained. The gravitational force can therefore in principle explore *all* the spatial dimensions of the theory, including the eleventh dimension of the bulk. And because the standard model particles of familiar experience are stuck to one of the D-branes, there is no restriction in principle on the size of the eleventh dimension. It could be small, wrapped up into a tiny cylinder. Or it could be large.

This was something of a revelation. It offers the possibility of using the difference between material particles and the standard model force-carriers (open strings) and the graviton (closed strings) to explain why gravity is so very different from the electromagnetic, weak and strong nuclear forces. The latter act on particles fixed to a D-brane. Gravity is free to roam.

Could this be an answer to the hierarchy problem? It didn't take superstring theorists too long to come up with some proposals. In 1998, theorists Nima Arkani-Hamed, Savas Dimopoulos and Gita Dvali (collectively referred to as ADD) devised a braneworld model in which the great gulf between the electro-weak mass-energy scale and the Planck mass-energy scale could be explained by the *dilution* of the force of gravity compared to other standard model forces.

The ADD model uses a single D-brane on which all the standard model particles can be found, and introduces hidden dimensions which are considerably larger than had been considered permissible in superstring theories to this point. Just a couple of large extra dimensions are needed to dilute the force of gravity sufficiently. Physicists (who are made of standard model particles) will therefore only ever measure gravity as it is manifested on the D-brane and will conclude that it is much weaker than the other forces of nature.

How large is 'large'? Well, contemporary experiments have verified Newton's classical inverse-square law of gravity down to millimetre dimensions. We would expect the effects of large hidden dimensions to show up as deviations from this law. This means that the hidden dimensions could be of the order of tenths of a millimetre, and we wouldn't know.

Now that's much, much larger than the Planck length.

A further braneworld model devised by American theorist Lisa Randall and Indian-born American Raman Sundrum uses two parallel D-branes, as in the original Hořava–Witten braneworld, but dilutes the effects of gravity not by using large hidden dimensions but by constructing a model in which the spacetime of the bulk is strongly warped by the energy it contains. On one brane, gravity is strong – as strong as the other standard model forces. But the warped spacetime between the branes dilutes gravity such that when it reaches the second brane, the force is considerably weaker. This second brane is where we are to be found, measuring a force of gravity that is now much weaker than the other forces, causing us to scratch our heads as we wonder how this can be.

I suspect that your reaction to braneworld scenarios such as these is really a matter of taste. Perhaps you're amazed by the possibility that there might be much more to our universe than meets the eye; that there might exist dimensions 'at right angles to reality' that we can't perceive but whose influence is manifested in the behaviour of those particles that we can observe. The revelation that there might be multidimensional branes, bulk and hidden dimensions – large, small or warped – might prompt more than one 'Oh wow!' moment.

But it doesn't do it for me, I'm afraid. I really can't read this stuff without thinking that I've accidentally picked up a Discworld novel, by the English author Terry Pratchett. Discworld is a fictional place not unlike earth, except that it is a flat disc, balanced on the backs of four elephants who in turn stand on the back of Great A'Tuin, a giant turtle which swims slowly through space. Magic is not unusual on Discworld, which has its own unique system of physics. I would speculate that people will be talking about Pratchett's Discworld long after the various braneworlds of M-theory have been quietly forgotten.

My problem is that branes and braneworld physics appear to be informed not by the practical necessities of empirical reality, but by imagination constrained only by the internal rules of an esoteric mathematics and an often rather vague connection with problems that theoretical physics beyond the standard model is supposed to be addressing. No amount of window-dressing can hide the simple fact that this is all *metaphysics*, not physics.

In his book *Facts and Mysteries in Elementary Particle Physics*, published in 2003, Martinus Veltman did not even wish to acknowledge supersymmetry and superstrings:

> The fact is that this is a book about physics, and this implies that the theoretical ideas discussed must be supported by experimental facts. Neither supersymmetry nor string theory satisfy this criterion. They are figments of the theoretical mind. To quote Pauli: they are not even wrong. They have no place here.[12]

Sheldon Glashow has wondered if superstring theory might be a more appropriate subject for mathematics departments, or even schools of divinity. 'How many angels can dance on the head of a pin?' he asked. 'How many dimensions are there in a compactified manifold, 30 powers of ten smaller than a pinhead?'[13]

The reality check

Am I being too harsh? After all, I made a big fuss in the opening chapter about the metaphysical nature of reality and the fact that we must be content with an empirical reality of things–as–they–appear or things–as–they–are–measured. Surely this means that any attempt to build a structure which goes beyond appearances and measurement is going to be metaphysical? And didn't I say that facts are in any case contaminated by theoretical concepts and that any path to a theory – no matter how speculative – is acceptable provided it yields a theory that works?

Yes, this is what I said. But if we accept the six Principles described in Chapter 1, then we must acknowledge that the determination of what is or what is not true in relation to a theory must be judged on the basis of the theory's empirical tests.

So, how do superstring theory and M-theory fare?

Let's set aside for the moment the fact that all the small print in the contract for supersymmetry must be carried over to the contract for superstring theory. This means 100-plus undetermined parameters, muons transforming into electrons and other problems. If superstrings are supersymmetric, then the supersymmetry must be broken. There are plenty of ideas, but no real consensus about how this is supposed to happen.

One key characteristic of superstring theories is that they require hidden dimensions. The fact that the extra dimensions are hidden or cannot be experienced directly doesn't necessarily mean that they exert no influence in our more familiar three-dimensional world. It is believed that telltale signals of extra dimensions might be observable in the form of so-called Kaluza–Klein (KK) particles. These particles are the mass projections into four dimensions of momenta that are carried by particles moving in the hidden dimensions.*

KK particles are a bit like the sparticles of supersymmetry. For every standard model particle able to travel in the hidden dimensions (i.e. provided they are not confined to a D-brane), there should exist KK particles with the same properties of spin and charge but with different masses. Large hidden dimensions imply light KK particles. In fact, there is a lightest Kaluza–Klein particle (LKP), equivalent to the LSP, which has also been considered as a possible dark matter candidate.

It will probably come as no surprise to learn that there are no hard and fast predictions for the masses of these particles, as indeed there are few, if any, predictions for anything else. Instead the theorists throw their general ideas – sparticles, KK particles – over the fence and hope that the experimentalists at CERN's LHC will take them seriously and/or get lucky. So far there is no evidence from the LHC for

---

* I'm reminded of the mice in Douglas Adams' *Hitchhiker's Guide to the Galaxy*. They are not mice, in fact. They are projections into our dimensions of super-intelligent, pan-dimensional beings.

anything more than what looks very much like the standard model Higgs boson.

The theorists have embraced superstring theory, the dualities of M-theory, branes and braneworlds like kids in a toyshop. They have been busy trying to come to terms with all the subtleties and interrelationships established within the mathematical frameworks of the theories. In other words, they have been trying to figure out how to play with the toys. It seems that they have found it easier to build braneworlds than to attempt to make predictions that can be properly tested.

As Gordon Kane admits:

> To be sure, the majority of research into string theory is not focused on how the theory connects to the real world; rather, most physicists are exploring questions at a more theoretical level. Such formal work is necessary, because … we need a deeper understanding to fully formulate the theory. Even the many theorists who are interested in how string theory connects to the real world don't typically think much about what it means to test the theory.[14]

Despite the beauty and tightness of the structure, the theorists still haven't learned enough about it to enable them to predict the properties of even a single standard model particle. Yau again:

> After some decades of exploring the Land of Calabi–Yau, string theorists and their math colleagues (even those equipped with the penetrating powers of geometric analysis) are finding it hard to get back home – to the realm of everyday physics (aka the standard model) – and, from there, to the physics that we know must lie beyond. If only it were as easy as closing our eyes, tapping our heels together and saying 'There's no place like home.' But then we'd miss out on all the fun.[15]

To date, thousands of papers and quite a few books have been published which explore the bizarre world of superstrings and M-theory. Most of these are concerned with the various esoterica of the theory and represent attempts to come to terms with its complex structure.

Nobody has yet found Dorothy's ruby slippers.

Richard Feynman was complaining about this situation as early as the late 1980s:

> I don't like that they're not calculating anything. I don't like that they don't check their ideas. I don't like that for anything that disagrees with an experiment, they cook up an explanation – a fix up to say 'Well, it still might be true.'[16]

The mis-selling* of the superstring programme

I don't actually mind that superstring and M-theory are more metaphysics than physics. More a modern fairy tale than a story of the real world. I don't mind that as many as 1,500 scientists have chosen to build careers in theoretical physics by chasing fantasies constructed on an arguably untenable structure of interconnected assumptions that now seem to have an ever-decreasing possibility of producing anything credible.** It is perhaps tragic that in pursuing the theory, the search for scientific truth has been abandoned, if not betrayed. But it's a free world and I could argue that within the physics community no real harm is done.

My problem is that this is *not* how the theory is presented to the wider public.

Thus, the jacket blurb for Lisa Randall's *Warped Passages* reads:

> Here she uses her experience at the cutting edge to reveal how the world is full of hidden, extra dimensions, far beyond our imaginations. She shows us how, in just a few years, we will be able to see them for the very first time – offering us a gateway into a whole new reality.[17]

---

* Mis-selling happens when a salesperson misrepresents or misleads an investor about the characteristics of a product or service (or theory), leaving out certain information or distorting the description to imply a need that doesn't exist.

** A recent estimate by Russian-born theorist Mikhail Shifman, an early proponent of SUSY, puts the number of high-energy physics theorists at somewhere between 2,500 and 3,000. He thinks of them as a 'lost generation', writing: 'During their careers many of them never worked on any issues beyond supersymmetry-based phenomenology or string theory. Given the crises (or, at least, huge question marks) we currently face in these two areas, there seems to be a serious problem in the community.' See M. Shifman, arXiv, pop-ph/1211.0004v3, 22 November 2012.

The jacket of Brian Greene's *The Fabric of the Cosmos* declares:

> Here [Greene] reveals a universe that is at once more surprising, exciting and stranger than any of us could have imagined ... It is a universe that exists in eleven dimensions, in which every single entity is composed of nothing more than tiny, vibrating pieces of string.[18]

Okay, an author has a lot less control than you might imagine over things like book title, cover design and jacket copy, and publishers understand that caveats, maybes, could-be's and other qualifications can put off readers and get in the way of selling books. Both Randall and Greene are somewhat more circumspect in the main text of their books.

But here's no less an authority than Stephen Hawking, who, with Leonard Mlodinow, writes in *The Grand Design*:

> M-theory is the unified theory Einstein was hoping to find. The fact that we human beings – who are ourselves mere collections of fundamental particles of nature – have been able to come this close to an understanding of the laws governing us and our universe is a great triumph. But perhaps the true miracle is that abstract considerations of logic lead to a unique theory that predicts and describes a vast universe full of the amazing variety that we see. If the theory is confirmed by observation, it will be the successful conclusion of a search going back more than 3,000 years. We will have found the grand design.[19]

Hawking has something of a reputation for this kind of thing. When he was appointed Lucasian Professor of Mathematics at Cambridge University (the chair once held by Newton and Dirac), he used his inaugural lecture to declare that the unified theory – presumably the one Einstein was hoping to find – was close at hand in the form of supergravity based on eight supersymmetries. That was over thirty years ago.

Do the authors of best-selling books about superstring theory actually believe in it? When asked directly at the Isaac Asimov Memorial Debate at the American Museum of Natural History in New York

City, which took place on 7 March 2011, Greene replied: 'If you asked me, "Do I believe in string theory?", my answer today would be the same as it was ten years ago: No.'[20]

He went on to explain that he only *believes* ideas that make testable predictions. But he insists that, despite the conspicuous absence of testable predictions from superstring theory, it remains one of the best bets for providing a unified theory.*

Greene's honesty is commendable. However, we might conclude that the continued publication of popular science books and the production of television documentaries that are perceived to portray superstring theory or M-theory as 'accepted' explanations of empirical reality (legitimate parts of the authorized version of reality) is misleading at best and at worst ethically questionable.

---

* Of course, this is a matter of opinion. Unfortunately I haven't been able in this book to tell you about alternative approaches to unifying general relativity and quantum theory. Unlike superstring theory, loop quantum gravity assumes no 'background' spacetime but rather generates the geometric framework of spacetime from within the theory, as does general relativity. The theory has made no immediately testable predictions, but clearly illustrates that the superstring programme is not the only game in town. Loop quantum gravity is the preserve of relativists, however, not particle theorists, and for this reason is less fashionable. I recommend Lee Smolin's popular books *Three Roads to Quantum Gravity* and *The Trouble with Physics*, and Carlo Rovelli's more specialist book *Quantum Gravity* (see the bibliography).

9

Gardeners of the Cosmic Landscape

Many Worlds and the Multiverse

Time and again the passion for understanding has led to the illusion that man is able to comprehend the objective world rationally by pure thought without any empirical foundations – in short, by metaphysics.

Albert Einstein[1]

SUSY, superstring theories and M-theory are relatively well-developed ingredients or candidates for a grand unified theory or a theory of everything. They seek to provide solutions for a number of the problems in the authorized version of reality: to offer a rationale for the masses of the three generations of elementary particles (in terms of string vibrations); to resolve the hierarchy problem; to provide dark matter candidates and finally to reconcile quantum theory with general relativity.

We can judge for ourselves how successful we think they have been at addressing these problems. They certainly cannot be considered as accepted components of the authorized version of reality. And there is a sense that, as more proton–proton collision data are collected at the LHC, at least some of the chickens may be about to come home to roost.

Of course this is disappointing. It would have been marvellous if the LHC had turned up signals that could have been interpreted as sparticles or mysterious Kaluza–Klein particles, shadowy projections of particles lurking in hidden dimensions. But although it is still relatively early days, the data appear to be telling us that this is probably not how nature works.

However, the fairy tales seem set to continue, and probably for some time to come. There appears to be just too much at stake. Too much effort has been expended for SUSY, superstring theories and M-theory to be given up lightly. And although the mathematical structures themselves might be tight, the network of largely unfounded assumptions on which the structures are based offer just too much freedom and flexibility. It seems that no amount of experimental data will be sufficient to close off all the 'work-arounds', the patches that allow the theorists to declare that, well, it might still be true.

But if we think superstring theorists have lost contact with reality and are indulging in the kind of metaphysics that Einstein warned us about in the quotation above, then we might need to prepare ourselves for a bit of a shock. The problems that SUSY, superstring theories and M-theory seek to address pale almost into insignificance compared with one of the most fundamental problems inherent in contemporary physical theory – the quantum measurement problem.

Seeking to resolve this problem has produced metaphysics on the very grandest of scales.

Von Neumann's 'projection postulate'

The first theorist carefully to articulate the mechanism (and hence the problem) of quantum measurement was the Hungarian mathematician John von Neumann. In his book *Mathematical Foundations of Quantum Mechanics*, first published in Berlin in 1932, he noted that the equations of quantum theory offer a perfectly respectable description of the time evolution of a quantum wavefunction which conforms entirely to our classical expectations. The wavefunction develops 'linearly', its properties and behaviour at each moment closely, smoothly and continuously connected with its properties and behaviour just a moment before. The time evolution is completely *deterministic*.

But when we intervene to make a measurement, we are obliged to abandon these equations. The measurement itself is like a quantum jump. It is 'discontinuous'. The properties of the measurement outcomes are not closely, smoothly and continuously connected to the initial wavefunction. The only connection between the initial wavefunction and the measurement outcomes is that the modulus squares of the amplitudes of the various components of the former can be used to determine the probabilities for the latter. The measurement is completely *indeterministic*. This is the 'collapse of the wavefunction'.

Von Neumann was also very clear that this collapse or process of 'projecting' the wavefunction into its final measurement state is not inherent in the equations of quantum theory. It has to be *postulated*, which is a fancy way of saying that it has to be assumed. Oh, and by the way, there is no experimental or observational evidence for this collapse per se. We just know that we start with a wavefunction which can be expressed as a superposition of the different possible outcomes and we end up with one – and only one – outcome.

There are, in general, three ways in which we can attempt to get around this assumption. We can try to eliminate it altogether by supplementing quantum theory with an elaborate scheme based on hidden variables. As the experiments described in Chapter 2 amply demonstrate, this scheme has to be very elaborate indeed. We know that hidden variables which reintroduce local reality – variables which establish localized properties and behaviours in an entangled quantum system, for example – are pretty convincingly ruled out by experiments that test Bell's inequality. We also know that the experiments designed to test Leggett's inequality tell us that 'crypto' non-local hidden variable theories in which we abandon the set-up assumption won't work either.

This leaves us with no choice but to embrace a full-blown non-local hidden variables theory.

Pilot waves

Such theories do exist, the best known being de Broglie–Bohm pilot wave theory, named for French theorist Louis de Broglie and American David Bohm. At great risk of oversimplifying, the de Broglie–Bohm theory assumes that the behaviour of completely localized (and therefore locally real) quantum particles is governed by a non-local field – the 'pilot wave' – which guides the particles along a path from their initial to their final states.

The particles follow entirely predetermined paths, but the pilot wave is sensitive to the measurement apparatus and its environment. Change the nature of the measurement by changing the orientation of a polarizing filter or opening a second slit, and the pilot wave field changes instantaneously in response. The particles then follow paths dictated by the new pilot wave field.

The de Broglie–Bohm theory has attracted a small but dedicated group of advocates, but it is not regarded as mainstream physics. To all intents and purposes, we have simply traded the collapse assumption for a bunch of further assumptions. Yes, we have avoided the collapse assumption and regained determinism – the fates of quantum particles are determined entirely by the operation of classical cause-and-effect principles. But we have also gained a pilot wave field which remains responsible for all the 'spooky' action-at-a-distance. And the end result is a theory that, by definition, predicts precisely the same results as quantum theory itself.

Einstein tended to dismiss this approach as 'too cheap'.[2]

Decoherence and the irreversible act of amplification

The second approach is to find ways to supplement quantum theory with a mechanism that makes the collapse rather more explicit. In this approach we recognize a basic, unassailable fact about nature – the quantum world of atomic and subatomic particles and the classical world of experience are fundamentally different. At some point we must cross a threshold; we must cross the point at which all the quantum weirdness – the superpositions, the phantom-like 'here' *and* 'there' behaviour – disappears. In the process of being amplified to scales that we can directly perceive, the superpositions are eliminated and the phantoms banished, and we finish up with completely separate and non-interacting states of 'here' *or* 'there'.

Is it therefore possible to arrange it so that Schrödinger's cat is never both alive and dead? Can we fix it so that the quantum superposition is collapsed and separated into non-interacting measurement outcomes long before it can be scaled up to cat-sized dimensions?

The simple truth is that we gain information about the microscopic quantum world only when we can amplify elementary quantum events and turn them into perceptible macroscopic signals, such as the deflection of a pointer against a scale. We never (but never) see superpositions of pointers (or cats). It stands to reason that the process of amplification must kill off this kind of behaviour before it gets to perceptible levels.

The physicist Dieter Zeh was one of the first to note that the interaction of a quantum wavefunction with a classical measuring apparatus and its environment will lead to rapid, irreversible decoupling or 'dephasing' of the components in a superposition, such that any interference terms are destroyed.

> Each state will now produce macroscopically correlated states: different images on the retina, different events in the brain, and different reactions of the observer. The different components represent two completely decoupled worlds. This decoupling describes exactly the ['collapse of the wavefunction']. As the 'other' component cannot be observed any more, it serves only to save the consistency of quantum theory.[3]

But why would this happen? In the process of amplification, the various components of the wavefunction become strongly coupled to the innumerable quantum states of the measuring apparatus and its environment. This coupling selects components that we will eventually recognize as measurement outcomes, and suppresses the interference. The process is referred to as *decoherence*.

We can think of decoherence as acting like a kind of quantum 'friction', but on a much faster timescale than classical friction. It recognizes that a wavefunction consisting of a superposition of different components is an extremely fragile thing.[*] Interactions with just a few photons or atoms can quickly result in a loss of phase coherence that we identify as a 'collapse'. This is fast but it is not instantaneous.

For example, it has been estimated that a large molecule with a radius of about a millionth (10^{-6}) of a centimetre moving through the air has a 'decoherence time' of the order of 10^{-30} seconds, meaning that the molecule is localized within an unimaginably short time and behaves to all intents and purposes as a classical object.[4] If we remove the air and observe the molecule in a vacuum, the estimated decoherence time increases to one hundredth of a femtosecond (10^{-17} seconds), which is getting large enough to be at least imaginable. Placing the

---

[*] This does not necessarily contradict the observation of the kinds of macroscopic superpositions described in Chapter 3, in which a wavefunction is formed in which billions of electrons flow in opposite directions around a superconducting ring. It just means that we have to work very hard to stop it from decohering.

molecule in intergalactic space, where it is exposed only to interactions with the cosmic microwave background radiation, increases the estimated decoherence time to 10^{12} seconds, meaning that a molecule formed in a quantum superposition state would remain in this state for a little under 32,000 years.

In contrast, a dust particle with a radius of a thousandth of a centimetre – a thousand times larger than the molecule – has a decoherence time in intergalactic space of about a microsecond (10^{-6} seconds). So, even where the possibility of interactions with the environment is reduced to its lowest, the dust particle will behave largely as a classical object.

The kinds of timescales over which decoherence is expected to occur for any meaningful example of a quantum system interacting with a classical measuring device suggest that it will be impossible to catch the system in the act of losing coherence. This all seems very reasonable, but we should remember that decoherence is an assumption: we have no direct observational evidence that it happens.

But does this really solve the measurement problem?

Decoherence eliminates the potentially embarrassing interference terms in a superposition. We are left with separate, non-interacting states that are statistical mixtures – different proportions of states that are 'up' or 'down', 'here' or 'there', 'alive' or 'dead'. We lose all the curious juxtapositions of the different possible outcomes (blends of 'up' and 'down', etc.). But decoherence provides no explanation for why *this* specific measurement should give *that* specific outcome. As John Bell has argued:

> The idea that elimination of coherence, in one way or another, implies the replacement of 'and' by 'or', is a very common one among solvers of the 'measurement problem'. It has always puzzled me.[5]

This is sometimes referred to as the 'problem of objectification'. Decoherence theory can eliminate all the superpositions and the interference, but we are still left to deal with quantum probability. We have no mechanism for determining which of the various outcomes – 'up'/'down', 'here'/'there', 'alive'/'dead' – we are actually going to get in the next measurement.

There are other theories that seek to make the collapse of the wavefunction explicit. But, of course, they all typically involve the replacement of the collapse assumption with a bunch of other assumptions which similarly have no basis in observation or experiment.

This leaves us with one last resort.

Everett's 'relative state' formulation of quantum theory

The third approach is to turn the quantum measurement problem completely on its head. If there is nothing in the structure of quantum theory to suggest that the collapse of the wavefunction actually happens, then why not simply leave it out? Ah, I hear you cry. We did that already and it led us to non-local hidden variables.

But there is another way of doing this that is astonishing in its simplicity and audacity. Let's do away with the collapse assumption and put our trust purely in quantum theory's deterministic equations. Let's not add *anything*.

Okay, I sense your confusion. If we don't add anything, then how can we possibly get from a smoothly and continuously evolving superposition of measurement possibilities to just one – and only one – measurement outcome? Easy. We note that as observers in this universe we detect just one – and only one – outcome. We assume that at the moment of measurement the universe splits into two separate, non-interacting 'branches'. In this branch of the universe you observe the result 'up', and you write this down in your laboratory notebook. But in another branch of the universe another you observes the result 'down'.

In one branch we run to fetch a bowl of milk for Schrödinger's very much alive and kicking cat. In another branch we ponder what to do with this very dead cat we found in a box. All the different measurement possibilities inherent in the wavefunction are actually realized. *But they're realized in different branches of the universe.*

As Swedish-American theorist Max Tegmark explained in the BBC *Horizon* programme mentioned in the Preface:

> I'm here right now but there are many, many different Maxes in parallel universes doing completely different things. Some branched off from this universe very recently and might look exactly the

same except they put on a different shirt. Other Maxes may have never moved to the US in the first place or never been born.[6]

I have always found it really rather incredible that the sheer stubbornness of the measurement problem could lead us here, to Hugh Everett III's 'relative state' formulation of quantum theory.

Everett was one of John Wheeler's graduate students at Princeton University. He began working on what was to become his 'relative state' theory in 1954, though it was to have a rather tortured birth. The theory was born, 'after a slosh or two of sherry',[7] out of a complete rejection of the Copenhagen interpretation and its insistence on a boundary between the microscopic quantum world and the classical macroscopic world of measurement (the world of pointers and cats).

The problem was that Wheeler revered Niels Bohr and regarded him as his mentor (as did many younger physicists who had passed through Bohr's Institute for Theoretical Physics in Copenhagen in their formative years). Wheeler was excited by Everett's work and encouraged him to submit it as a PhD thesis. But he insisted that Everett tone down his language, eliminating his anti-Copenhagen rhetoric and all talk of a 'splitting' or 'branching' universe.

Everett was reluctant, but did as he was told. He was awarded his doctorate in 1957 and summarized his ideas in an article heavily influenced by Wheeler which was published later that year. Wheeler published a companion article in the same journal extolling the virtues of Everett's approach.

It made little difference. Bohr and his colleagues in Copenhagen accused Everett of using inappropriate language. Everett visited Bohr in 1959 in an attempt to move the debate forward, but neither man was prepared to change his position. By this time the disillusioned Everett had in any case left academia to join the Pentagon's Weapons System Evaluation Group. He went on to become a multimillionaire.

Despite Wheeler's attempts to massage the Everett theory into some form of acceptability, there was no escaping the theory's implications, nor its foundation in metaphysics. It seemed like the work of a crackpot. Wheeler ultimately came to reject it, declaring: '… its infinitely many unobservable worlds make a heavy load of metaphysical baggage'.[8]

Many worlds

And so Everett's interpretation of quantum theory languished for a decade. But in the early 1970s it would be resurrected and given a whole new lease of life.

The reason is perhaps not all that hard to understand. The Copenhagen interpretation's insistence on the distinction between microscopic quantum system and classical macroscopic measurer or observer is fine in practice (if not in principle) when we're trying to deal with routine 'everyday' measurements in the laboratory. But early theories of quantum cosmology, culminating in the development of the ΛCDM model, tell us that there was a time when the *entire universe* was a quantum system.

When American theorist Bryce DeWitt, who had also been a student of Wheeler, developed an early version of a theory of quantum gravity featuring a 'wavefunction of the universe', this heralded the beginnings of a quantum cosmology. There was no escaping the inadequacy of the Copenhagen interpretation in dealing with such a wavefunction. If we assume that everything there is is 'inside' the universe, then there can be no measuring device or observer sitting outside whose purpose is to collapse the wavefunction of the universe and make it 'real'.

DeWitt, for one, was convinced that there could be no place for a special or privileged 'observer' in quantum theory. In 1973, together with his student Neill Graham, he popularized Everett's approach as the 'many worlds' interpretation of quantum theory, publishing Everett's original (unedited) Princeton PhD thesis in a book alongside a series of companion articles.

In this context, the different 'worlds' are the different branches of the universe that split apart when a measurement is made. According to this interpretation, the observer is unaware that the universe has split. The observer records the single result 'up'. She scratches her head and concludes that the wavefunction has somehow mysteriously collapsed. She is unaware of her parallel self, who is also scratching her head and concluding that the wavefunction has collapsed to give the single result 'down'.

If this were really happening, how come we remain unaware of it? Wouldn't we retain some sense that the universe has split? The answer given by early proponents of the Everett formulation is that the laws of quantum theory simply do not allow us to make this kind of observation.

In a footnote added to the proofs of his 1957 paper, Everett accepted the challenge that a universe that splits every time we make a quantum measurement appears to contradict common experience (and sense). However, he went on to note that when Copernicus first suggested that the earth revolves around the sun (and not the other way around), this view was initially criticized on the grounds that nobody had ever directly experienced the motion of the earth through space. Our inability to sense that the earth is moving was eventually explained by Galileo's theory of inertia. Likewise, Everett argued, our inability to sense a splitting of the universe is explained by quantum physics.

If the act of quantum measurement has no special place in the many worlds interpretation, then there is no reason to define measurement as being distinct from any process involving a quantum transition between initial and final states. Now, we would be safe to assume that there have been a great many quantum transitions since the big bang origin of the universe. Each transition will have therefore split the universe into as many worlds as there were contributions in all the different quantum superpositions. DeWitt estimated that there must by now be more than a googol (10^{100}) worlds.

As Wheeler himself remarked, the many worlds interpretation is cheap on assumptions, but expensive with universes.

As the different worlds in which the different outcomes are realized are completely disconnected from each other, there is a sense in which Everett's formulation anticipated the emergence of decoherence theory. Indeed, variants of the many worlds interpretation that have been developed since DeWitt resurrected it have explicitly incorporated decoherence, avoiding the problem of objectification by assuming that different outcomes are realized in different worlds.

The original Everett conception of a universe which splits into multiple copies has undergone a number of reinterpretations. Oxford theorist David Deutsch argued that it is wrong to think of the universe splitting with each quantum interaction, and proposed instead that there exist a possibly infinite number of parallel worlds or parallel universes, among which the different outcomes are somehow partitioned. These parallel universes have become collectively known as the 'multiverse'.[*]

Are we really supposed to take this seriously?

It is certainly true that an increasing number of theorists are sufficiently perplexed by the quantum measurement problem that they are willing to embrace many worlds, despite all its metaphysical baggage. At a scientific workshop on the interpretation of quantum theory held in August 1997, the participants conducted an informal poll. Of the 48 votes recorded, 13 (or 27 per cent) were cast in favour of the Copenhagen interpretation. The second most popular choice, attracting eight votes (17 per cent) was the many worlds interpretation. Tegmark reported:

> Although the poll was highly informal and unscientific (several people voted more than once, many abstained, etc.), it nonetheless indicated a rather striking shift in opinion compared to the old days when the Copenhagen interpretation reigned supreme. Perhaps most striking of all is that the many worlds interpretation … proposed by Everett in 1957 but virtually unnoticed for about a decade, has survived 25 years of fierce criticism and occasional ridicule to become the number one challenger to the leading orthodoxy…[9]

Before we go any further, I think we should probably remind ourselves of precisely what it is we're dealing with here. The many worlds interpretation singularly avoids assumptions regarding the collapse of the wavefunction and does not seek to supplement or complete quantum theory in any way. Rather, it takes the framework provided by quantum theory's deterministic equations and insists that this is all there is.

But the consequence of this approach is that we are led inexorably to the biggest assumption of all:

---

[*] Strictly speaking, the term 'universe' means 'everything there is', but it quickly gets confusing if we refer to lots of parallel universes as 'the universe'. The logic that prevails here is that we consider our known universe of visible stars, galaxies, dark matter and dark energy as merely one among many such universes that exist alongside ours in parallel. We call this collection of different parallel universes the 'multiverse'.

> **The Many Worlds Assumption.** *The different possible outcomes of a quantum measurement or the different possible final states of a quantum transition are all assumed to be realized, but in different equally real worlds which either split from one another or exist in parallel.*

The reality check

Attempts to rehabilitate the many worlds interpretation by devising versions that appear to carry less metaphysical baggage have been broadly frustrated. Despite the reasonableness of these versions and their use of less colourful language, the measurement problem remains particularly stubborn. We might at this stage be inclined to think that, after all, there's no real alternative to the many worlds interpretation's more extreme logic. Perhaps we should just embrace it and move on?

But before we do, it's time for another reality check.

The first thing we should note is that the many worlds interpretation is not a single, consistent theory that carries the support of all who declare themselves 'Everettians'. It is not even a single, consistent interpretation. It's more correct to say that there is a loose-knit group of theorists who buy into the idea that the quantum measurement problem can be solved by invoking the many worlds assumption. But each has rather different ideas about precisely how this assumption should be made and how the practical mechanics should be handled. As Cambridge theorist Adrian Kent recently noted:

> Everettian ideas have been around for 50 years, and influential for at least the past 30. Yet there has never been a consensus among theoretical physicists either that an Everettian account of quantum theory can be made precise and made to work, or that the Everettian programme has been comprehensively refuted.[10]

One of the single biggest challenges to the acceptance of the many worlds interpretation lies in the way it handles probability. In conventional quantum theory in which we assume wavefunction collapse, we use the modulus squares of the amplitudes of the different components in the wavefunction to determine the probabilities that these components will give specific measurement outcomes. Although we might have our suspicions about what happens to those components that 'disappear' when the wavefunction collapses, we know that if we continue to repeat the measurement on identically prepared systems, then all the outcomes will be realized with frequencies related to their quantum probabilities.

Let's use an example. Suppose I form a superposition state which consists of both 'up' and 'down' (it doesn't matter what actual properties I'm measuring). I take 0.866 times the 'up' wavefunction and 0.500 times the 'down' wavefunction and add them together. I now perform a measurement on this superposition to determine if the state is either 'up' or 'down'. I cannot predict which result I will get for any specific measurement, but I can determine that the probability of getting the result 'up' is given by 0.866^2, or 0.75 (75 per cent), and the probability of getting the result 'down' is 0.500^2, or 0.25 (25 per cent).* This means that in 100 repeated experiments, I would expect to get 'up' 75 times and 'down' 25 times. And this is precisely what happens in practice.

How does this translate to the many worlds scenario? The many worlds assumption suggests that for each measurement, the universe splits or partitions such that each 'up' result in one world is matched by a 'down' result in a parallel world, and vice versa. If in this world we record the sequence 'up', 'up', 'down', 'up' (consistent with the 75:25 probability ratio of conventional quantum theory), then is it the case that the other we in another world record the sequence 'down', 'down', 'up', 'down'? If so, then this second sequence is clearly no longer consistent with the probabilities we would calculate based on the original superposition.

Things are obviously not quite so simple.

Perhaps each repeated measurement splits or partitions the sequence among more and more worlds? But how then do we recover a sequence that is consistent with our experience in the one world we do observe? Should the probabilities instead be applied somehow to the worlds

---

* Strictly speaking, the probabilities are given by $|0.866|^2$ and $|0.500|^2$. We use the modulus-squares because the coefficients could be complex (they could include i, the square root of -1).

themselves? What would this mean for parallel worlds that are meant to be equally real? Should we extend the logic to include an infinite number of parallel worlds?

In *The Hidden Reality*, Brian Greene acknowledges the probability problem in a footnote:

> So from the standpoint of observers (copies of the experimenter) the vast majority would see spin-ups and spin-downs in a ratio that does not agree with the quantum mechanical predictions ... In some sense, then ... (the vast majority of copies of the experimenter) need to be considered as 'nonexistent'. The challenge lies in understanding what, if anything, that means.[11]

As we dig deeper into the problem, we realize that the seductive simplicity that the many worlds interpretation seemed to offer at first sight is, in fact, an illusion.

Various solutions to the problem of probability in the many worlds interpretation have been advanced, and Everett himself was convinced that he had shown in his PhD thesis how the quantum mechanical probabilities can be recovered. But it is fair to say that this is the subject of ongoing debate, even within the community of Everettians.

Eternal inflation

The multiverse interpretation that we have so far considered has been invoked as a potential solution to the quantum measurement problem. Its purpose is to free the wavefunction from the collapse assumption so that it can be applied to describe the universe that we observe. If this was as far as it went, then perhaps the multiverse of many worlds would not provoke more than an occasional research paper, a chapter or two in contemporary texts on quantum theory and the odd dry academic conference.

But there's obviously more to it than this.

We have seen how cosmic inflation provides solutions to the horizon and flatness problems in big bang cosmology. However, inflation is more a technique than a single cut and dried theory, and it may come as no surprise to discover that there are many ways in which it can be applied. One alternative approach was developed in 1983 by

221

Russian theorist Alexander Vilenkin, and was further elaborated in 1986 by Russian-born theorist Andrei Linde.

One of the problems with the current ΛCDM model of big bang cosmology is that it leaves us with quite a few unanswered (and possibly unanswerable) questions. Perhaps one of the most perturbing is the fine-tuning problem. In order to understand why the universe we inhabit has the structure and the physical laws that it has, we need to wind the clock back to the very earliest moments of its existence. But it is precisely here that our theories break down or run beyond their domain of applicability.

But now here's a thought. We can try to determine the special circumstances that prevailed during the very earliest moments of the big bang, circumstances that shaped the structure and physical laws that we observe. Alternatively, we can question just how 'special' these circumstances actually were. Just as Copernicus argued that the earth is not the centre of the universe; just as modern astronomy argues that our planet orbits an unexceptional star in an unexceptional galaxy in a vast cosmos, could we argue that our entire universe is no more than a relatively unexceptional bit of inflated spacetime in an unimaginably vaster multiverse?

If inflation as a technique is freed from the constraint that it should apply uniquely to our universe, then all sorts of interesting things become possible. Linde discovered that it was possible to construct theories on a rather grander scale:

> There is no need for quantum gravity effects, phase transitions, supercooling or even the standard assumption that the universe originally was hot. One just considers all possible kinds and values of scalar [i.e. inflaton] fields in the early universe and then checks to see if any of them leads to inflation. Those places where inflation does not occur remain small. Those domains where inflation takes place become exponentially large and dominate the total volume of the universe. Because the scalar fields can take arbitrary values in the early universe, I called this scenario chaotic inflation.[12]

The kind of model developed originally by Vilenkin and Linde is now more commonly referred to as *eternal inflation*. In this model, our universe is merely one of countless 'bubbles' of inflated spacetime, triggered by quantum fluctuations in a vast inflaton field (or fields) driven by competition between the decay of the field's energy and the exponential growth of energy in pockets of inflation. In certain models the competition is unequal, and the bubbles proliferate like a virus, or like the bubbles in a bottle of champagne when the cork is extracted.

The bubbles form a multiverse much like holes in an ever-inflating piece of Swiss cheese. This is, of course, an imperfect analogy, as it is the holes or bubbles themselves (rather than the cheese) that account for much of the spacetime volume. Such an 'inflationary multiverse' could be essentially eternal, with no beginning or end.

In the inflationary multiverse, anything is possible. The essential randomness of the quantum fluctuations that trigger bubbles of inflation imply a continuum of universes with different physical laws (different cosmological constants, for example).

Let's reserve judgement on this idea for now, and simply note that we're dealing here with another assumption.

The Inflationary Multiverse Assumption. *Certain inflationary cosmological models describe a multiverse consisting of bubbles of inflating spacetime triggered by quantum fluctuations in a vast inflaton field. Our universe may be a relatively unexceptional bubble in this multiverse.*

We need to be clear that the multiverse of the many worlds interpretation and the multiverse of eternal inflation are necessarily different. They originate from within different theoretical structures and have been proposed for very different reasons. But, as the saying goes: in for a penny, in for a pound. If we're going so far as to invoke the idea of a multiverse, why not simplify things by assuming that the bubble universes demanded by eternal inflation are, in fact, the many worlds demanded by quantum theory?

In May 2011, American theorists Raphael Bousso and Leonard Susskind posted a paper on the arXiv pre-print archive in which they state:

> In both the many-worlds interpretation of quantum mechanics and the multiverse of eternal inflation the world is viewed as an

unbounded collection of parallel universes. A view that has been expressed in the past by both of us is that there is no need to add an additional layer of parallelism to the multiverse in order to interpret quantum mechanics. To put it succinctly, the many-worlds and the multiverse are the same thing.[13]

Breathtaking.

But we're still not quite there. Joining the many worlds interpretation to the inflationary multiverse cannot explain the current fascination with multiverse theories. To understand this fascination, we must return once again to superstring theory. As Alan Guth explained in 2007:

Until recently, the idea of eternal inflation was viewed by most physicists as an oddity, of interest only to a small subset of cosmologists who were afraid to deal with concepts that make real contact with observation. The role of eternal inflation in scientific thinking, however, was greatly boosted by the realization that string theory has no preferred vacuum ...[14]

The cosmic landscape

In *The Hidden Reality*, Brian Greene wrote of his early experiences with Calabi–Yau shapes, the manifolds which are used to roll up and hide away the six extra spatial dimensions demanded by superstring theory:

When I started working on string theory, back in the mid-1980s, there were only a handful of known Calabi–Yau shapes, so one could imagine studying each, looking for a match to known physics. My doctoral dissertation was one of the earliest steps in this direction. A few years later, when I was a postdoctoral fellow (working for the Yau of Calabi–Yau), the number of Calabi–Yau shapes had grown to a few thousand, which presented more of a challenge to exhaustive analysis – but that's what graduate students are for. As time passed, however, the pages of the Calabi–Yau catalog continued to multiply ... they have now grown more numerous than grains of sand on a beach. Every beach.[15]

Greene wasn't kidding. In 2003, theorists Shamir Kachru, Renata Kallosh, Andrei Linde and Sandip Trivedi worked out the number of different Calabi–Yau shapes that are theoretically possible. This number is determined by the number of 'holes' each shape can possess, up to a theoretical maximum of about five hundred. There are ten different possible configurations for each hole. This gives a maximum of 10^{500} different possible Calabi–Yau shapes.

The precise shape of the Calabi–Yau manifold determines the nature of the superstring vibrations that are possible. It thus determines the physical constants, the laws of physics and the spectrum of particles that will prevail. In other words, the shape determines the type of universe that will result. The figure 10^{500} therefore refers to the number of different possible *types* of universe, not the total number of possible universes. This is what Guth meant when he talked about string theory having no preferred vacuum.

I believe there was a time in the history of physics when this kind of result would have been taken as evidence that a theoretical programme had failed. We could conclude that 10^{500} different possible Calabi–Yau shapes with no compelling physical reason to select the one shape that uniquely describes our universe – and hence describes the laws and the particles that we actually observe – leave us with nowhere to go. Time to go back to the drawing board.

Except, of course, we now have eternal inflation and the inflationary multiverse.

Far from this vast multiplicity of possible Calabi–Yau shapes being seen as evidence for the failure of the superstring programme, it is instead used to bolster the idea that what the theory is describing is actually a multiverse. Greene again:

> The idea is that when inflationary cosmology and string theory are melded, the process of eternal inflation sprinkles string theory's 10^{500} possible forms for the extra dimensions per bubble universe – providing a cosmological framework that realizes all possibilities. By this reasoning, we live in that bubble whose extra dimensions yield a universe, cosmological constant and all, that's hospitable to our form of life and whose properties agree with observations.[16]

Each bubble in the inflationary multiverse represents a universe that may be much like the one we inhabit, or it might be subtly different, or it might be vastly different. It doesn't take a great leap of the imagination to suggest that each bubble is characterized by the Calabi–Yau shape that governs its extra spatial dimensions.

Leonard Susskind calls it the cosmic landscape. He is at pains to explain that the landscape of possibilities afforded by the 10^{500} Calabi–Yau shapes is not 'real'. It is rather a list of all the different possible designs that universes could possess. However, he is unequivocal on the reality of the multiverse: 'The pocket [i.e. bubble] universes that fill it are actual existing places, not hypothetical possibilities.'[17]

Let's log it as another in what is proving to be a long series of interconnected assumptions.

The Landscape Assumption. *The 10^{500} different possible ways of compactifying the six extra spatial dimensions of superstring theory represent different possible types of universe that may prevail within the inflationary multiverse.*

The universe next door

The inflationary multiverse provides a mathematical metaphor for a cosmos in which countless bubbles of spacetime are constantly inflating like balloons. Any region of spacetime devoid of content but for the inflaton field is potentially unstable and susceptible to quantum fluctuations which may trigger inflation. When combined with superstring theory's demand for extra 'hidden' dimensions, we conclude that the bubbles are characterized by different Calabi–Yau shapes, giving rise to universes with different physical constants, laws and particles.

Of course, we have already encountered something very similar to this. The D-branes of M-theory represent extended string-like objects that can accommodate whole universes with different physical constants, laws and particles. Indeed, the Hořava–Witten and Randall–Sundrum braneworld scenarios are examples of cosmological models involving parallel universes. In the latter, the two branes describe universes in which gravity acts very differently. In one universe it is strong. The warped spacetime of the bulk then dilutes the gravitational force so that when we experience it in our braneworld, is it considerably weakened.

It doesn't require a great leap of imagination to propose that there may be many more than just two braneworlds 'out there'. If the structure of each brane is governed by the Calabi–Yau shape that hides its extra dimensions, then M-theory would suggest that the multiverse is actually one of parallel braneworlds.

On the surface, this is yet another take on the multiverse idea. But the brane multiverse is rather different. The many worlds interpretation and the inflationary multiverse leave us with a rather static picture. The parallel universes are almost by definition detached one from another and they do not interact. But in M-theory, branes are dynamic objects. They move around (in the bulk). And they can interact.

What happens when brane universes collide? They don't just pass through each other like ghostly ships in the night. M-theory suggests that a collision between braneworlds could be very violent.

In 2001, Princeton theorists Paul Steinhardt and Justin Khoury, working with Burt Ovrut at the University of Pennsylvania and Neil Turok, then at Cambridge University, used heterotic M-theory to study the effects of colliding braneworlds. They concluded that, under the right circumstances, the collision of a 'visible' 3-brane containing four large spacetime dimensions and standard model particles with an 'invisible' 3-brane moving slowly through the bulk could in principle trigger a cataclysmic release of energy. A small proportion of the kinetic energy (the energy of motion) of the invisible brane is converted into hot radiation, which bathes the matter in the visible brane.

The hot radiation is later recognized as a hot big bang. It accelerates the expansion of spacetime without requiring the mechanism of inflation as used in the ΛCDM model and eternal inflation. The collision resets the visible brane's time clock, and, to all intents and purposes, the universe we observe today is born. The theorists wrote:

> Instead of starting from a cosmic singularity with infinite temperature, as in conventional big bang cosmology, the hot, expanding universe in our scenario starts its cosmic evolution at a finite temperature. We refer to our proposal as the 'ekpyrotic

universe', a term drawn from the Stoic model of cosmic evolution in which the universe is consumed by fire at regular intervals and reconstituted out of this fire, a conflagration called ekpyrosis. Here, the universe as we know it is made (and, perhaps, has been remade) through a conflagration ignited by collisions between branes along a hidden fifth dimension.[18]

As far as I can tell, the term 'ekpyrotic universe' didn't catch on. Some have referred to the scenario instead as the *big splat*, which is something of a misnomer, as the branes don't so much splat as bounce off each other.

I should make clear that Steinhardt and Turok and their colleagues do not promote their work as an argument in favour of a brane multiverse. Rather, they have developed a cosmology which seeks to explain our universe in terms of an eternal cycle: two branes collide and bounce apart, but never separate far enough to escape their mutual gravitational influence (which is manifested in our universe as dark energy). Some trillions of years later the branes collide again, the visible universe is 'reset' and the cycle repeats.

In this colliding braneworld scenario, the pattern of temperature variations observed in the CMB radiation is the result of quantum fluctuations in both braneworlds at the point of collision. The branes are not perfectly 'flat'. Quantum uncertainty means that the braneworlds come into contact first in places where the amplitudes of fluctuations in the direction of the bulk dimension are high. These cause 'hotspots' – places in the visible brane where the radiation temperature is higher than the average.

Steinhardt, Turok and their colleagues have continued to develop this cyclic cosmology, which they now refer to as a 'phoenix universe', championing it as a viable alternative to cosmologies based on inflation. In 2008, Steinhardt and Princeton theorist Jean-Luc Lehners adapted the model to prevent spacetime collapsing into an amalgam of black holes. In this revised model, much of the universe is destroyed but a small volume (of the order of a cubic centimetre) survives the ekpyrotic phase (hence the term 'phoenix'). 'As tiny as a cubic centimetre may seem, it is enough to produce a flat, smooth region a cycle from now at least as large as the region we currently observe. In this way, dark energy, the big crunch and the big bang all work together so that the phoenix forever arises from the ashes, crunch after crunch after crunch.'[19]

The reality check (II)

There are other kinds of multiverse theory that unfortunately we don't have the space to consider here. In *The Hidden Reality*, Greene identifies nine different types, including many worlds, the inflationary multiverse, the brane and cyclic multiverses and the cosmic landscape.

This is surely an extraordinary situation. At great risk of repeating myself, I want to summarize briefly the steps that have brought us here.

The current authorized version of reality consists of a collection of partially connected theoretical structures – quantum theory, the standard model of particle physics, the special and general theories of relativity and the ΛCDM model of big bang cosmology. There remain many significant problems with these structures and we know this can't be the final story. But there is little guidance available from experiment and observation. There are no big signs pointing us towards possible solutions.

Into this vacuum a number of new theoretical structures have been introduced. They offer potential solutions for some of the problems but they are inevitably motivated by the theorists' instincts and their mathematical and personal preferences. The theories are constructed on a network of assumptions that we are obliged to accept at face value.

So, superstring theory is founded on the assumption that elementary particles can be represented by vibrations in one-dimensional filaments of energy. To this we add the assumption that there exists a fundamental spacetime symmetry between fermions and bosons. As the theory then demands a total of nine spatial dimensions, we assume that six dimensions are compactified into a Calabi–Yau space. There is no experimental or observational evidence for any of these assumptions.

In a separate series of developments, the quantum measurement problem is resolved by assuming that all the different possible outcomes of a measurement are realized in different equally real worlds which either split from one another or exist in parallel. Eternal inflation is a characteristic of certain cosmological models. These models describe a multiverse consisting of bubbles of inflating spacetime triggered by quantum fluctuations in a vast inflaton field. We assume our universe is a relatively unexceptional bubble in this multiverse. We further assume that each of these universes can be characterized by one of the 10^{500} different possible ways of compactifying the six extra spatial dimensions demanded by superstring theory. In an alternative scenario, the hot big bang origin of our own universe could be the result of a collision between two braneworlds. There is no experimental or observational evidence for any of these assumptions.

Let's just check to see if we've understood this correctly. We live in a multiverse, 'surrounded' by parallel universes that by definition we cannot experience directly. We can never verify the existence of these universes, and must look instead for evidence that betrays their existence indirectly in the physics of our own universe.

Of course, there is no evidence in the physics of the authorized version of reality, so we must look to the physics of superstrings or M-theory. And look! The fact that there is no preferred choice of Calabi–Yau shape from the 10^{500} different possibilities is taken to imply that our universe is far from unique. There must be many, many other kinds of universe. Quod erat demonstrandum.

Justifying a multiverse theory because it is implied by superstring theory might just be acceptable if superstring theory was an already accepted description of the universe we experience. But it's not. In fact, one of the things standing in the way of superstring theory's acceptance is its inability to predict anything about our own universe. And this is partly because, with 10^{500} different Calabi–Yau spaces to choose from, virtually anything and everything is possible. There's nothing in the theory that allows us to pick out the space that describes our universe uniquely.

The multiverse theory is justified by superstring theory but superstring theory cannot be proved because we live in a multiverse.

We've not just crossed a line here. We're so far over it that, to quote Matt LeBlanc as Joey in an episode of *Friends*, the line is a dot to us. To be fair, some of the most recent popular presentations of multiverse theories have included sections titled 'Is this science?' or something similar. Some authors argue that the multiverse theory might be testable through the observation of legacy effects in the CMB from a big bang triggered by colliding branes. Others argue that the multiverse must be science because the cosmic landscape is derived from superstring theory and superstring theory is science. Frankly, these look very much like attempts to clutch at straws, and none are convincing.

Reference to the six Principles described in Chapter 1 would lead us to conclude that theories constructed on many worlds and the multiverse are not testable *in principle*. To those who would argue that there are genuine instances in which parallel universes could impress themselves on the physics we observe in our universe, I would respectfully suggest that physicists will struggle to demonstrate that any such evidence is free from ambiguity. I honestly can't believe that it will be beyond the wit of theorists to find alternative physical explanations for such instances based on theories requiring one – and only one – universe.

Of course, this is just an opinion.

Dealing with the fallout

As you may have gathered by now, I'm rather exasperated by the relentless pursuit of SUSY, superstring theory and M-theory, and the widening gap between increasingly implausible theoretical speculations and the practicalities of experiment and observation. But I also believe that the willingness to embrace multiverse theories – what Peter Woit refers to as 'multiverse mania' – is doing serious damage to the cause of science itself.*

Firstly, there is mis-selling of the kind I highlighted for the superstring programme. Multiverse theories have divided the physics community and the debate is growing in vocal intensity. Although there are many notable advocates, such as Susskind, Tegmark, Greene (a relatively recent convert) and Martin Rees (Britain's Astronomer Royal), there are also notable dissidents. These include David Gross, Steinhardt and Lee Smolin. Despite this disagreement, the multiverse is still paraded in the popular science media as an accepted body of scientific theory.

Thus, in an 'ultimate guide' to the multiverse published in the respected UK popular science weekly *New Scientist* in November 2011, science writer Robert Adler opens with:

---

* See http://www.math.columbia.edu/~woit/wordpress/?cat=10.

Whether we are searching the cosmos or probing the subatomic realm, our most successful theories lead to the inescapable conclusion that our universe is just a speck in a vast ocean of universes.

He does not explain how 'success' is defined. Nowhere in this article can I find any reference to the fact that this is not a universally accepted theory, or that multiverse theories are not actually 'inescapable'. Or that a debate is raging about whether this is even science. On the contrary, he continues:

> Three decades after the concept was born, many researchers now agree that at least some kinds of multiverse stand on firm theoretical foundations, yield predictions that astronomers and particle physicists can test, and help explain why our universe is the way it is.[20]

But how many is 'many'? And, should you be in any doubt, please be reassured that *no* multiverse theory of any kind can explain why our universe is the way it is.

The second area of concern relates to the really rather obvious fact that multiverse theories represent an enormous cop-out. By arguing that all the different measurement outcomes and all the different possible universes are realized in the multiverse, theorists duck the challenge to understand why we experience these outcomes in our universe and why our universe has the physical constants, laws and spectrum of particles that it has. Greene himself acknowledges this problem in *The Hidden Reality*:

> By invoking a multiverse, science could weaken the impetus to clarify particular mysteries, even though some of those mysteries might be ripe for standard, nonmultiverse explanations. When all that was really called for was harder work and deeper thinking, we might instead fail to resist the lure of multiverse temptation and prematurely abandon conventional approaches.[21]

The multiverse answer is that all possible universes exist in parallel and we just happen to find ourselves in one that supports our form of life.*
Of course, this is no answer at all. Greene goes on to defend the multiverse approach, arguing that to abandon it because it could be a blind alley is equally dangerous.

Finally, and most importantly, we must be concerned about the implications of multiverse theories for the future development of science itself. The multiverse theorists know that they are on weak ground regarding the Testability Principle, and rather than admit that their theories are not science, they argue instead that the rules of science must be adapted to accommodate this kind of metaphysical speculation.

They want to change the very *definition* of science. This is a very slippery slope.

---

* We will further explore this kind of reasoning in Chapter 11.

10

Source Code of the Cosmos

Quantum Information, Black Holes and the Holographic Principle

> *The physicist has to limit himself very severely: he must content himself with describing the most simple events that can be brought within the domain of our experience; all events of a more complex order are beyond the power of the human intellect to reconstruct with the subtle accuracy and logical perfection the theoretical physicist demands.*
>
> Albert Einstein[1]

The Reality Principle introduced in Chapter 1 argues that reality consists of things-in-themselves which are, by definition, inaccessible to us and of which we can never hope to gain knowledge. Science is the process by which we continue to refine our knowledge of the things-as-they-appear or the things-as-they-are-measured, from which we can *infer* what reality is 'really' like if we first assume that reality has objective, independent existence.

As we have no way of knowing what such an independent reality is actually like, this leaves us with some considerable freedom to speculate.

The reductionists among us might be tempted simply to conclude that reality must be ultimately composed of irreducible bits or fields of 'stuff', however we choose to define these. These bits or fields impress themselves on our empirical reality of observation and measurement in ways that are hopefully logical and accessible to human reason. Whatever reality is, this stuff manifests or projects itself to produce effects which we interpret as quarks and leptons and force particles, and so on.

But what if, instead, there is no 'stuff'?[*] After all, Einstein set some remarkable precedents in the early twentieth century. In the special and general theories of relativity he showed that space and time are not absolute, they are relative. They depend on the things that exist within them and, it can be argued, they have no existence independently of these things.

The Higgs mechanism turns mass into a secondary quality, like colour. We begin to get the sense that, though tangible, mass is an effect – it is the result of things interacting with the Higgs field – rather than something primary, something that exists independently of such interactions.

And then the experimental violation of Leggett's inequality tells us that what we call 'properties' – for example, the spin orientations of quarks and leptons – cannot be considered to be inherently pre-existing attributes of these things. They are instead products of the process of measurement – they are the 'projections' (if that's the right word) of the things-in-themselves into our empirical reality of measurement.

The story of modern physics is really the story of the not-so-gradual undermining of our naïve ideas about reality. It tells of the erosion of our broadly common-sense, taken-for-granted notions and their replacement by uncertainty and doubt. Can we expect this erosion of confidence to continue indefinitely?

What if we stripped away all the things we call 'properties', the things that define what it means for a quantum particle to be present in a specific quantum state, the things that determine how the particle is going to behave. What would be left?

There are at least two possible answers to this question. The first is that by stripping away all the empirical dressing – energy, mass, spin, space, time, etc. – what we are left with is abstract *mathematics*. A second answer is that we could consider the irreducible 'stuff' of the universe (or multiverse) to be *information*.

---

[*] Or, as Neo pondered in a scene in *The Matrix*, perhaps there is no spoon.

The Mathematical Universe Hypothesis

The principal exponent of the notion that the core description of everything in the universe might be essentially mathematical is MIT theorist Max Tegmark. Now, theorists have always tended to wax rather lyrical about what Eugene Wigner called the 'unreasonable effectiveness of mathematics in the natural sciences'.[2] The creative processes involved in fashioning theories out of the often arcane and esoteric concepts, rules and language of mathematics are arguably equivalent in many ways to art. And just as great art seems to tell us vivid truths about human existence, so perhaps great maths relates deep truths of physical existence.

In this hypothesis, however, Tegmark argues that the effectiveness of mathematics is not at all unreasonable, or even surprising.

First, some groundwork. The hypothesis is premised on what Tegmark calls the External Reality Hypothesis (ERH). There exists an external physical reality that is completely independent of human beings.

This is hardly radical. The ERH defines what it means to be a realist, though as I have explained, there can be no observational or experimental verification for such a hypothesis. The assumption of an independent reality is really an act of faith, or, if this is a tad too theological for your taste, an act of metaphysics.

What is radical is what Tegmark concludes from the ERH: 'The Mathematical Universe Hypothesis (MUH): Our external physical reality is a mathematical structure.'[3]

What exactly is this supposed to mean?

Imagine strolling barefoot along a beach late one afternoon. The sun is arcing towards the distant horizon, and we're killing time picking up interesting-looking shells that happen to cross our path. You pick up three shells and add these to the four you already have in your pocket. How many shells do you now have? Easy. You know that three plus four equals seven.

A mathematical structure consists of a set of abstract entities and the relations that can be established to exist between these. In our beachcombing example, the abstract entities are integer numbers and the relations consist of addition, subtraction, etc. These are the relations of ordinary algebra.

The idea of numbers and the relationships between them is so commonplace that we often don't think of it as 'abstract' at all. But, of course, in our real-world beachcombing scenario, there appear to be only shells. The idea that there are numbers of shells and that these can be added or subtracted in a logical fashion is an *additional* structure that is arguably inherent in our empirical reality of shells. In the MUH, Tegmark argues that when we strip away reality's empirical dressing (the shells, me, you, the beach and everything else), what we are left with is a universe defined solely by the abstract entities and their relations – the numbers and the algebra.

In fact, he goes further. Our tendency is to deploy different mathematical structures as appropriate in an attempt to describe our single (empirical) universe. Tegmark argues that the structure *is* the universe. Every structure that can be conceived therefore describes a different (parallel) universe.

This is not an entirely new vision. The ancient Greek philosopher Plato argued that the independent reality of things-in-themselves consists of perfect, unchanging 'forms'. The 'forms' are abstract concepts or ideas such as 'goodness' and 'beauty', accessible to us only through our powers of reason. Our empirical reality is then an imperfect or corrupt projection of these forms into our world of perception and experience. If we think of the 'forms' as abstract mathematical structures, then Tegmark's MUH is a kind of radical Platonism. As Tegmark himself explained in an interview with Adam Frank for *Discover* magazine:

> Well, Galileo and Wigner and lots of other scientists would argue that abstract mathematics 'describes' reality. Plato would say that mathematics exists somewhere out there as an ideal reality. I am working in between. I have this sort of crazy-sounding idea that the reason why mathematics is so effective at describing reality is that it is reality. That is the mathematical universe hypothesis: Mathematical things actually exist, and they are actually physical reality.[4]

This kind of logic obviously begs all sorts of questions. One possible counter-argument is that mathematical structures are not independently existing things. Rather, they are actually *human inventions*. They are systems of logic and reasoning with concepts, rules and language that we have devised and which we find particularly powerful when used to describe our physical world. By associating mathematics with the human mind in this way, we conclude that in a universe with no minds there can be no mathematics – mathematics is not an independently existing thing that minds have a knack of 'discovering'.

You might be inclined to conclude that this is really all just some kind of philosophical nit-picking, and I would be tempted to agree with you. But Tegmark claims that the MUH is *testable*. If mathematical structures exist independently of the things they are used to describe, then physicists can expect to continue to uncover more and more mathematical regularities. And if our universe is but one in what Tegmark calls the 'Level IV' multiverse, then we can test this by determining the statistical likelihood of a universe described by the mathematical structure that prevails compared with universes described by other structures.

I, for one, don't find this very convincing. You can come to your own conclusions. Despite his claims of testability, Tegmark himself seems to acknowledge that it is really all philosophical speculation. In an apparent throwaway remark towards the end of the interview with *Discover*, he comments that his wife, a respected cosmologist, 'makes fun of me for my philosophical "bananas stuff", but we try not to talk about it too much'.

Now that sounds like good advice.

Quantum information

The second possibility is that the basic stuff of the universe is *information*. So how is this supposed to work?

This logic is driven from the observation that the physical world appears to consist of *opposites*. We have positive and negative, spin-up and spin-down, vertical and horizontal, left and right, particle and anti-particle, and so on. Once again, if we strip away the empirical dressing, such as charge, spin, etc., what we are left with is a fundamental 'oppositeness'. An elementary 'on' and 'off'.

Or, alternatively, an elementary '0' and '1'.*

---

* Yes, yes, I know. Quarks and leptons come in *three* generations. Quark colour comes in *three* varieties – red, green and blue. Boy, you can be really pedantic at times!

In one of the simplest mathematical structures we can devise (or discover, depending on your point of view), the abstract entities are 'bits' which have one of only two possible values, 0 or 1. As most readers will be aware, these are the basic – so-called 'binary' – units of information used in all computer processes.

But now here's a twist. Classical bits have the values 0 *or* 1. Their values are one or the other. They cannot be added together in mysterious ways to make superpositions of 0 *and* 1. However, if we form our bits from quantum particles such as photons or electrons, then curious superpositions of 0 and 1 become perfectly possible. Such 'quantum bits' are often referred to as 'qubits'. Because we can form superpositions of qubits, the processing of such quantum information works very differently compared with the processing of classical information.

Suppose we have a system consisting of just two classical bits. The bits have values 0 or 1, so there are four (or $2^2 = 2 \times 2$) different possible 'bit strings'. Both bits could have the value 0, giving the bit string 00. There are two possibilities for the situation where one bit has the value 0 and one has the value 1 – 01 and 10. Finally, if both bits have the value 1, the bit string is 11.

If we extend this logic to three classical bits, then we anticipate eight (or $2^3 = 2 \times 2 \times 2$) different possible bit strings: 000, 100, 010, 001, 110, 101, 011 and 111. We could go on, but the process quickly becomes tedious and in any case you get the point. A system of n classical bits gives 2^n different possible bit strings.

But a system of two or three classical bits will form only *one* of these bit strings at a time. In a system consisting of two or three qubits, we can form *superpositions* of all these different possible combinations. The physical state of the superposition is determined by the amplitudes of the wavefunctions of each qubit combination, subject to the restriction that the modulus squares of the amplitudes sum to 1 (measurement can give one, and only one, bit string). This means that the state of a superposition of n qubits is described by 2^n amplitude factors.

Here's where it gets interesting. If we apply a computational process to a classical bit, then the value of that bit may change – the bit string changes from one possibility to another. For example, in a system with two bits, the string may change from 00 to 10. But applying a computational process to a qubit superposition changes all 2^n components of the superposition *simultaneously*. An input superposition yields an output superposition. This is important. When we apply a computation to an input in a classical computer, we get a linear output. In a quantum computer we get an *exponential* amount of computation in the same amount of time.

But, hang on. What about the collapse of the wavefunction? Isn't it the case that when we make a measurement to discover what the qubit string actually is, we lose all the other components of the superposition? Yes, this is true. However, by exploiting quantum interference effects between different computational paths, we can fix it so that the probability of observing the correct bit string (i.e. the string that represents the logically correct result of the computation) is enhanced and all the other bit strings are suppressed.

Exponentially scaling up the output of a computation sounds vaguely like a good thing to do, but if we're going to discard most of the possible results then where's the benefit? But the fact is that we don't 'discard' the other results. We set up the input superposition so that the computation proceeds exponentially and the amplitudes of the components in the output superposition are 'concentrated' around the logically correct result.

It's difficult to comprehend just what this means without some examples. The prospects for quantum computation were set out by Oxford theorist David Deutsch in 1985, but it took almost ten years for computer scientists to develop algorithms based on its mechanics.

In 1994, American mathematician Peter Shor devised a quantum algorithm for finding the prime factors of large numbers. Three years later, Indian-born American computer scientist Lov Grover developed a quantum algorithm that produces database search results within a square root of the amount of time required by a classical computer.

These examples may not sound particularly earth-shattering, but don't be misled. The cryptographic systems used for most internet-based financial transactions (such as the RSA algorithm)* are founded on the simple fact that factoring large integer numbers requires a vast amount of computing power, regarded as virtually impossible with conventional computers. For example, it has been estimated that a

---

* Named for Ronald Rivest, Adi Shamir and Leonard Adelman. The RSA algorithm is a form of public-key encryption.

network of a million conventional computers would require over a million years to factorize a 250-digit number. Yet this feat could in principle be performed in a matter of minutes using Shor's algorithm on a single quantum computer.[5]

Excitement is building. There have been many recent reports of practical, though small, laboratory-scale quantum computers. Entangled quantum states and superpositions are extremely fragile and can decohere very quickly, so any system relying on the constant establishment (and collapsing) of entangled states has to be operated in a very carefully managed environment.

In February 2012, researchers at IBM announced significant technological advances which bring them 'very close to the minimum requirements for a full-scale quantum computing system as determined by the world-wide research community'.[6]

Stay tuned.

Information and entropy

Advances in quantum computing make the subject of quantum information both fascinating and important. But other than notions that the universe could be considered to be one vast quantum computer, they don't get us any closer to the idea that information might be the ultimate reality.

Indeed, we might be inclined to dismiss this idea for the same reason we might have dismissed the MUH. We could argue that the concept of information, like mathematics, is an abstraction based on the fundamental properties of the empirical entities of the material universe. Quantum particles have properties that we can interpret in terms of information. But, we might conclude, quantum information cannot exist without quantum properties. Which came first? The properties or the information? The chicken or the egg?

But there is one relationship that might cause us to at least pause and think before concluding that only physical properties can shape physical reality. It lends some credibility to the notion that information itself might be considered as a *physical* thing.

This is the relationship between information and entropy.

Entropy is a thermodynamic quantity that we tend to interpret as the amount of 'disorder' in a system. For example, as a block of ice melts, it transforms into a more disordered, liquid form. As liquid water is heated to steam, it transforms into an even more disordered, gaseous form. The measured entropy of water increases as water transforms from solid to liquid to gas.

The second law of thermodynamics claims that in a spontaneous change, entropy always increases. If we take a substance – such as air – contained in a closed system, prevented from exchanging energy with the outside world, then the entropy of the air will increase spontaneously and inexorably to a maximum as it reaches equilibrium with its surroundings. It seems intuitively obvious that the oxygen and nitrogen molecules and trace atoms of inert gas that make up the air will not all huddle together in one corner of the container. Instead, the air spreads out to give a uniform air pressure. This is the state with maximum entropy.

Unlike other thermodynamics quantities, such as heat or work, entropy has always seemed a bit mysterious. The second law ties entropy to the 'arrow of time', the experience that despite being able to move freely in three spatial dimensions – forward–back, left–right, up–down – we appear obliged to follow time in only one direction – forwards. Suppose we watch as a smashed cocktail glass spontaneously reassembles itself, refills with Singapore sling and flies upwards through the air to return to a guest's fingers. We would quickly conclude that we're watching a film of these events playing *backwards* in time.

Austrian physicist Ludwig Boltzmann established that entropy and the second law are essentially statistical in nature. Molecules of air might be injected into one corner of the container, but they soon discover that this system has many more microscopic physical states available than the small number accessible to them huddled in the corner.* Statistically speaking, there are many more states in which the air molecules move through all the space available in the container than there are states in which the molecules group together.

Another way of putting this is that the macroscopic state with a uniform average distribution of molecular positions and speeds is the most *probable*, simply because there are so many more microscopic

---

* By 'microscopic', I mean a state defined by the individual positions and speeds of the molecules. A macroscopic state is then defined by *averaging* over all the different microscopic state possibilities.

states that contribute to the average. The air molecules expand in the container from a less probable to a more probable macroscopic state, and the entropy increases. Boltzmann discovered that the entropy is proportional to the logarithm of the number of possible microscopic states that the system can have. The higher the number of these states, the higher the probability of the macroscopic state that results.

Note that this is all about statistical probabilities. There is in principle nothing preventing all the air molecules in my study from suddenly rushing into one corner of the room, causing me to die from asphyxiation. It's just that this macroscopic state of the air molecules is very, very improbable.

There's yet another way of thinking about all this. Suppose we wanted to keep track of the positions and velocities of molecules in a sample of water. This is obviously a lot easier to do if the water is in the form of ice, as the molecules form a reasonably regular and predictable array with fixed positions. But as we heat the ice and convert it eventually to steam, we lose the ability to keep track of these molecules. The molecules are all still present, and we can use statistics to give us some notion of their average speeds, but we can no longer tell where every molecule is, where it's going or how fast.

Now imagine we could apply a similar logic to one of the great soliloquies from Shakespeare's *Macbeth*. From Act V, scene v, we have:

> She should have died hereafter;
> There would have been a time for such a word.
> Tomorrow, and tomorrow, and tomorrow,
> Creeps in this petty pace from day to day,
> To the last syllable of recorded time;
> And all our yesterdays have lighted fools
> The way to dusty death. Out, out, brief candle!
> Life's but a walking shadow, a poor player
> That struts and frets his hour upon the stage
> And then is heard no more. It is a tale
> Told by an idiot, full of sound and fury
> Signifying nothing.[7]

Let's suppose we can 'heat' this soliloquy. At first, the passage melts and the words lose their places in the structure – 'And syllable but shadow a frets sound all ...' Eventually, the words come apart and transform into a soup of individual letters – 's', 't', 'A', 'n', 'e' ... But the letters of the English alphabet can be coded as a series of bit strings.* With further heating, the bit strings come apart to produce a random 'steam' of bits – '0', '0', '1', '0', '1' ...

All the resonance and meaning in the soliloquy – all the *information* it contained – has not exactly been lost in this process. After all, we still have all the bits. But our ability to recover the information has become extremely difficult. Our ignorance has increased. It would take an enormous amount of effort to reconstruct the soliloquy from the now scrambled bits, just as it would take an awful lot of work to reconstruct the cocktail glass from all the shards picked up from the floor. If the information isn't lost, then it has certainly become almost irretrievably 'hidden' (or, alternatively, our ignorance of the soliloquy has become very stubborn).

What this suggests is that there is a deep relationship between information and entropy.

In 1948, American mathematician and engineer Claude Shannon developed an early but very powerful form of information theory. Shannon worked at Bell Laboratories in New Jersey, the prestigious research establishment of American Telephone and Telegraph (AT&T) and Western Electric (it is now the research and development subsidiary of Alcatel-Lucent). He was interested in the efficiency of information transfer via communications channels such as telegraphy, and he found that 'information' as a concept could be expressed as the logarithm of the inverse of the probability of the value of a random variable used to communicate the information.

What does this mean? Suppose we agree a simple code to communicate a particular state of affairs, such as 'my plane took off on time'. We code this event as '1'. The alternative, 'my plane did not take off on time', is coded as '0'. Knowing airline punctuality as you do, you expect to receive the message '0' with 85 per cent probability. If you do indeed receive the message '0', then Shannon's formula

---

* In the American Standard Code for Information Interchange (ASCII) system, the letter 's' (for example), transcribes as the bit string 1110011.

implies that its information content is very low.* But if you receive the message '1', then the information content is high. In this context, information is a measure of the 'unexpectedness' of the message, or the extent to which you're surprised by it.

So, the entropy of Macbeth's soliloquy is low,** and its information content is high (the particular sequence of words is of low probability and therefore unexpected or 'surprising', as well as very moving). Heat the soliloquy into a steam of bits. The entropy increases and the information content is, if not lost, then hidden very, very deeply.

In 1961, this kind of logic led IBM physicist Rolf Landauer to declare that 'information is physical'. He was particularly interested in the processing of information in a computer. He concluded that when information is erased during a computation, it is actually dumped into the environment surrounding the processor, adding to the entropy. This increase in entropy results in an increase in temperature: the environment surrounding the processor heats up. Anyone who has ever run a complex computation on their laptop computer will have noticed how, after a short while, the computer starts to get uncomfortably hot.

Landauer's famous statement requires some careful interpretation, but it's enough for now to note the direct connection between the processing of information and physical quantities such as entropy and temperature. It seems that 'information' is not an abstract concept invented by the human mind. It is a real, physical thing with real, physical consequences.

Now we're coming to it. The second law of thermodynamics insists that in a spontaneous change, entropy will always increase or information will always be diluted or degraded and so 'hidden'. So what then happens when we throw stuff – a volume of steam, for

---

* A high probability (p) has a consequently low inverse probability ($1/p$) and the logarithm of this will be small.

** The entropy of Shakespeare? I worked as a post-doctoral researcher at Oxford University in the early 1980s. In those days, analysing experimental data typically required a Fortran program running on a mainframe computer. I'd occasionally get irritated by the fact that some students were hogging the teletype machines in the computer centre. Bafflingly, they seemed to be typing the entire works of Shakespeare into a computer program. I learned later that they were using the program to determine the entropy of Shakespeare.

example – into a black hole? By definition, when the material crosses the black hole's event horizon, there's no coming back. If we dispose of a lot of high-entropy material in a black hole, this seems to imply that the net entropy of the universe has somehow reduced. Material with high entropy has disappeared from the universe. The change is spontaneous, so this appears to contradict the second law, which says that entropy can never decrease.

And if the entropy of the universe has reduced, this implies that its information content has somehow increased.

It is estimated that the total information in the universe is of the order of 10^{120} bits.[8] If we regard information to be an elementary, physical constituent of the universe, then this implies that, like energy, it can be neither created nor destroyed. How, then, can throwing high-entropy material into a black hole increase the information content of the universe?

Black holes and the second law

Gravity might be the weakest of nature's forces, but it is ultimately irresistible. Gravity binds together and compresses clouds of gas drifting in space. Compressing clouds of gas with sufficient mass sparks fusion reactions at their cores, and stars are born. The pressure of radiation released by the fusion reactions holds back further compression, and the star enters a period of relative stability.

However, as the fuel is expended, the force of gravity grips tighter. For any mass greater than about 1.4 $M_\odot$ (1.4 solar masses),* the force of gravity is ultimately irresistible. It crushes the body of matter into the obscurity of a black hole, a name coined by John Wheeler.

For a time it was thought that black holes would indeed invalidate the second law. The only way to preserve the law would be to ensure that the entropy of the material that was consumed by a black hole was somehow transferred to the black hole itself.

But does it make any sense to think of a black hole as something that has entropy?

---

* This is known as the Chandrasekhar limit, named for Indian physicist Subrahmanyan Chandrasekhar.

In the late 1960s, young Cambridge University physicist Stephen Hawking produced a series of papers on black hole physics in collaboration with mathematician Roger Penrose, then at Birkbeck College in London. General relativity, they claimed, predicted that at the heart of a black hole there beats a singularity, a region of infinite density and spacetime curvature where the laws of physics break down. Of course, what goes on in the region of a singularity is completely hidden from observation by the black hole's event horizon, a fact that Penrose elevated to the status of a principle, which he called the cosmic censorship hypothesis.

Working with Canadian Werner Israel, Australian Brandon Carter and British physicist David Robinson, Hawking demonstrated that, in terms of the mathematics needed to describe them, holes are surprisingly simple. Their properties and behaviour depend only on their mass, angular momentum and electric charge, a conjecture called the 'no hair' theorem. In this context, 'hair' means all other kinds of information apart from mass, angular momentum and electric charge. Beyond these basic properties, a black hole is featureless – it has no hair. All other kinds of information are judged to be lost behind the black hole's event horizon.

In a moment of inspiration one night in November 1970, Hawking realized that the properties of the event horizon meant that this could never shrink – the surface area of a black hole (meaning the area bounded by the event horizon) could in principle never decrease. If a black hole consumes an amount of material, then its surface area increases by a proportional amount.

We do a mental double-take. Isn't there one other well-known physical property that in a spontaneous change can never decrease? So, could there be a connection between the surface area of a black hole and its entropy?

In his 1972 Princeton PhD thesis, Israeli theorist Jacob Bekenstein (another of Wheeler's students) claimed precisely this. He identified the surface area of a black hole with its entropy. So, when a black hole consumes some high-entropy material, its surface area increases (as Hawking had observed), and this indicates that its entropy increases too, in accordance with the second law.

He was shouted down from all sides.

Hawking was irritated. While this appeared to offer a neat solution, it dragged along with it a number of implications which he felt

Bekenstein hadn't properly addressed. For one thing, a body with entropy also has to have a temperature. And a body with a temperature has to emit radiation. This made no sense at all. How could a black hole, with properties and behaviour determined only by its mass, angular momentum and electric charge, possess a temperature and *emit* radiation?

A black hole is supposed to be 'black'. Right?

Hawking radiation

A few years later, Hawking set out to refute Bekenstein's hypothesis. Lacking a fully fledged quantum theory of gravity, he chose to approach the problem using an essentially classical general relativistic description of the black hole itself, and applied quantum field theory to the curved spacetime around the event horizon. What he found was quite shocking.

As he later explained:

> However, when I did the calculation, I found, to my surprise and annoyance, that even nonrotating black holes should apparently create and emit particles at a steady rate. At first I thought that this emission indicated that one of the approximations I had used was not valid. I was afraid that if Bekenstein found out about it, he would use it as a further argument to support his ideas about the entropy of black holes, which I still did not like.[9]

Hawking found that within the constraints imposed by Heisenberg's uncertainty relation, virtual particle–anti-particle pairs are produced in the curved spacetime near the black hole's event horizon. The particles are produced with zero net energy. This means that one particle in the pair may possess positive energy and the other negative energy.

Under normal circumstances, the pair would quickly annihilate. But if the negative energy particle were to be drawn into the black hole before it can be annihilated, it can acquire positive energy and become a 'real' particle. Another way of looking at this is to think of the negative energy particle as having negative mass (from $E = mc^2$). As the negative mass particle falls into the back hole, it gains mass and becomes a real particle.

The positive energy particle created in the virtual pair may then escape, to all intents and purposes appearing as though it has been emitted by the black hole.

Bekenstein had been right all along. The entropy of a black hole is proportional to its surface area.* A black hole does have a temperature.** Black holes ain't so black, after all. They emit what has since become known as *Hawking radiation*. There is no spontaneous reduction in entropy; no spontaneous increase in the universe's information content.

But there *was* more trouble. Hawking showed that as negative energy (negative mass) particles spill through the event horizon, the black hole must lose mass overall and its surface area must therefore *decrease*. This apparent reduction in entropy is more than compensated for by the entropy of the emitted Hawking radiation. So, having demonstrated that the second law holds even when material is consumed by a black hole, there was no immediate threat of violating the law as a result of emitting Hawking radiation.

As the black hole emits radiation, its surface area decreases. Consequently, its temperature increases, as does the rate of emission. The black hole eventually 'evaporates', disappearing altogether in an explosion.

It's important to hold on here to a simple fact. At no time in this evaporation process has anything come *out* of the black hole. Although its surface area has shrunk, and its temperature and the 'glow' of Hawking radiation has increased, the whole process is driven by particles falling *into* the black hole.

And this is the problem. Think of everything that goes past the black hole's event horizon in terms of so many bits of information. What happens to all these bits when the black hole evaporates?

Hawking was unequivocal:

---

* Hawking demonstrated that the entropy of a black hole is proportional to one quarter of the black hole surface area, measured in units defined by the Planck area (the square of the Planck length), about 0.26 billionths of a trillionth of a trillionth of a trillionth of a trillionth of a trillionth of a square metre (2.6×10^{-70} m^2).

** Although this *is* very small. A black hole of 1 $M_\odot$ is expected to have a temperature just 60 billionths of a degree above absolute zero. It would actually absorb more cosmic microwave background radiation than it emits.

When a black hole evaporates, the trapped bits of information disappear from our universe. Information isn't scrambled. It is irreversibly, and eternally, obliterated.[10]

This was *not* good. If information is indeed physical, then it should not be possible to destroy it in this way. But there was an even more immediate worry. In the absence of measurement, the physical state of a quantum object as it evolves in time is determined by the information carried in its wavefunction. It is a key postulate of quantum theory that this kind of information connects the future with the past and so must be conserved.

If, as Hawking was now arguing, black holes can destroy such information, the entire basis of quantum theory is threatened.

The black hole war

It was called the black hole 'information paradox'.

Theorists Gerard 't Hooft and Leonard Susskind heard about Hawking's challenge directly from Hawking himself at a small private scientific conference in San Francisco in 1981. It was tantamount to a declaration of war.

Hawking is, arguably, a relativist. Both 't Hooft and Susskind are elementary particle theorists, for whom quantum theory – and the conservation of information it demands – is sacrosanct. Hawking just *had* to be wrong. But neither could provide an instant refutation.

Over the next twelve years, there were sporadic skirmishes, but battle was properly joined in 1993 at another conference organized at the Institute for Theoretical Physics* at the University of California at Santa Barbara. Susskind led the charge. At the start of his lecture, he announced: 'I don't care if you agree with what I say. I only want you to remember that I said it.'[11]

What Susskind had to say seemed vaguely mad. There were not all that many options, and like Sherlock Holmes, Susskind figured that

---

* Now named the Kavli Institute for Theoretical Physics, in recognition of a substantial donation to the institute from Norwegian-born Fred Kavli, a physicist, inventor, entrepreneur and philanthropist.

when he had eliminated all the impossible options, what remained, however improbable, must be the truth.

If the scrambled bits of information were not to be lost for ever inside the evaporating black hole, then either they were somehow preserved on its surface, to be eventually emitted in the form of Hawking radiation, or they were preserved in some kind of remnant left behind after the black hole had evaporated completely. The latter seemed unlikely, so Susskind pitched for the former.

He argued that the processes involved in transferring information to a black hole must be subject to a curious kind of complementarity.

To an observer watching from a safe distance, high-entropy material (for some reason in these scenarios this is nearly always an unfortunate astronaut) approaches the event horizon. The astronaut encounters what Susskind called the 'stretched horizon', a hot, Planck-length-thick layer surrounding the event horizon from which the Hawking radiation escapes, much as the very top of the earth's atmosphere evaporates into space. Here, he meets his inevitable fate. He is reduced to scrambled bits of information.

But, Susskind argued, the bits of information formerly known as the astronaut stay trapped on the surface of the black hole, each occupying a 'cell' with an area equal to four times the Planck area. The bits are eventually emitted as Hawking radiation, which is what the distant observer sees. Although reconstituting the astronaut would be an extremely difficult (though not completely impossible) task, no bits are lost.

But this doesn't seem to square with what we think we know about black holes. Susskind explained that there is another, complementary, perspective. The astronaut himself observes something quite different. From his perspective, he passes through the stretched horizon and the event horizon without noticing anything particularly unusual. He passes the point of no return, possibly without even realizing it. He is eventually torn apart by gravitational tidal forces and destroyed by the singularity. The bits of information formerly known as the astronaut are irretrievably lost.

How can this make any sense? Susskind argued, much as Bohr had done in the 1920s, that despite appearances, these two very different perspectives are not actually contradictory. They are complementary. With the help of Canadian theorist Don Page, he was able to show that the two perspectives are mutually exclusive, like the wave and particle perspectives of conventional quantum theory. We can observe what happens from a safe distance or we can join the astronaut on his journey through the event horizon. But we can't do both.

Page and Susskind were able to prove that it is not possible to recover information from the emitted Hawking radiation and then plunge with this into the black hole in search of the same information that the astronaut has carried into the interior. This turns out to be broadly analogous to showing that an electron cannot have both wave and particle properties simultaneously. By the time the information has been recovered from the Hawking radiation and transported into the interior of the black hole, the same information carried by the astronaut has already been destroyed by the singularity. The bits can't coexist.

A straw poll of the theorists gathered in Santa Barbara suggested that Susskind had won this round. More than half of those present agreed that information is not lost inside a black hole, but is recovered in the Hawking radiation that it emits.

Susskind wasn't satisfied, however. He realized he needed a firmer mathematical basis for his notion of black hole complementarity. This was provided by 't Hooft a year later, and championed by Susskind through the use of a startling visual metaphor.

The holographic principle

Here's a clue. If I want to work out how much information I can pack into the British Library in London, I would probably start by working out how many shelves I can get into the *volume* of space that the building contains. So how come all this talk about information and black holes has all been about the black hole's surface *area*?

I guess the simple answer is that we really have no clue about what goes on inside a black hole's event horizon, and so we can say nothing really meaningful about its volume. In other words, volume is a measure that is by definition *interior*. To resolve the black hole information paradox, we need to work with the only measure that is still accessible to us, the measure that defines the black hole but remains firmly *exterior* – its area.

The really rather intriguing thing about the next step, however, is that it offered a generalization that takes us a long, long way from black hole physics. Area, it turns out, is fundamentally connected with information in a way that has nothing to do with black holes.

In 1994, Susskind visited 't Hooft at the University of Utrecht in the Netherlands. 't Hooft told him of a paper he had written some months before. As he explained his most recent work, Susskind realized what was really going on. On his way back to California, he began working on what was to become known as the *holographic principle*.

Simply put, this principle says that the information content of a bounded volume of space – for example, a black hole bounded by its event horizon – is equivalent to the information content held on the boundary. More generally, the information contained in an n-dimensional space is equivalent* to the information on its $(n-1)$-dimensional boundary surface. As 't Hooft had done in his paper, Susskind now compared this to the way a hologram works:

> On the second floor of the Stanford [University] physics department, there used to be a display of a hologram. Light reflecting off a two-dimensional film with a random pattern of tiny dark and light spots would focus in space and form a floating three-dimensional image of a very sexy young woman who would wink at you as you walked past.[12]

According to the maths, this is a general principle, not something that is specific to black holes. Susskind went on to speculate that the information content of the entire universe – in other words, everything in the universe, including me, you and Max Tegmark – is actually a low-energy projection of the information 'encoded' on the universe's cosmic horizon.

In this interpretation of the holographic principle, our three-dimensional world is an illusion. It is really a hologram, like the three-dimensional image of the sexy young woman who winks at you as you walk by. Reality is actually information stored on the boundary of the universe.

---

* For the purists, the information in the n-dimensional space is *isomorphic* with information contained in its $(n-1)$-dimensional bounding surface.

This is Plato's allegory of the cave in reverse. In that story, the prisoner perceived reality as a two-dimensional shadow projection of a three-dimensional 'real' world. The holographic principle says that our perceived three-dimensional reality is actually a projection from a two-dimensional hologram 'painted' on the boundary of the universe.

This is where we find the source code of the cosmos.

Holography and superstring theory

The holographic principle might well have remained an interesting curiosity in information theory and a fascinating slice of metaphysics. But in 1998, Argentinian theorist Juan Maldacena announced a powerful new result. He showed that the physics described by a Type IIB superstring theory within an n-dimensional bulk spacetime is entirely equivalent to the physics described by low-energy quantum field theory applied to its $(n-1)$-dimensional boundary. This was a whole new superstring duality.

What makes this duality extraordinary is that low-energy quantum field theory does not include gravity. Yet its dual superstring theory does.

In his paper, Maldacena did not make explicit the connection between his result and the holographic principle. But just a few months later Witten posted a paper elaborating Maldacena's result. On seeing Witten's paper, Susskind understood that the black hole war had finally been won:

> Quantum field theory is a special case of quantum mechanics, and information in quantum mechanics can never be destroyed. Whatever else Maldacena and Witten had done, they had proved beyond any shadow of a doubt that information would never be lost behind a black hole horizon. The string theorists would understand this immediately; the relativists would take longer. But the war was over.[13]

Indeed, Witten used this new duality to show that a black hole in the bulk spacetime is equivalent to a relatively mundane hot 'soup' of elementary particles, such as gluons, on the boundary surface.

There are caveats. Maldacena had considered a model consisting of a stack of D-branes sandwiched together. D-branes are places where the ends of open strings 'stick'. The open strings can wander all over the surface of a D-brane, but they can't escape it. Closed strings, on the other hand, are free to wander through the bulk.

The stack of D-branes form a slab. Open strings can now wander over the different layers of the slab, and their ends may be located on a single layer or on different layers. When only low-energy configurations (low-mass strings) are considered, closed strings (and hence gravity) are eliminated and the model can be represented by a quantum field theory such as QCD, in which the open strings are gluons.

Maldacena then changed perspective. We can consider the D-branes as surfaces on which open strings move about, but we can also consider them as physical entities in their own right, with their own energy and mass. Stack enough D-branes together and spacetime starts to warp, just as it does in the presence of any mass. Add more D-branes and we cross a threshold. We form a *black brane*. The curved spacetime created by the black brane has some singular properties. It is a so-called anti-de Sitter space (or AdS).

In 1917, Dutch physicist Willem de Sitter solved Einstein's gravitational field equations for a model universe empty of matter in which spacetime expands exponentially. We can think of such a universe as consisting only of dark energy, producing a positive cosmological constant. It therefore has positive curvature. As on the surface of a sphere, the angles of a triangle drawn in de Sitter space will add up to more than $180°$.

In an anti-de Sitter space, the cosmological constant is negative, the spacetime curvature is negative and the angles of a triangle add up to less than $180°$.* This is a hyperbolic universe shaped more like a saddle. Inject matter into an anti-de Sitter universe and the curvature of spacetime causes it to be pushed away from the boundary and drawn towards the centre, even in the absence of a conventional gravitational field.

Closed strings moving through the bulk of this anti-de Sitter space are no different in principle from closed strings moving through the different D-brane layers in the first perspective. But closed strings that

---

* Recall that Euclidean geometry, in which the angles of a triangle add up to precisely $180°$, is based on the assumption of a 'flat' spacetime.

wander close to the boundary defined by the black brane appear very different. The curvature of anti–de Sitter space in the region of the boundary pushes against the closed strings, such that they appear to lose energy.

This was Maldacena's insight. These two apparently very different perspectives must describe the same physics. Open strings moving through a stack of D-branes (described equivalently in terms of a quantum field theory without gravity) must yield the same physics as low-energy closed strings (described by a Type IIB superstring theory with gravity) near the boundary of an anti–de Sitter space. The type of quantum field theory used is referred to as conformal field theory, or CFT. This relationship that Maldacena had discovered is then called a CFT/AdS duality.

The caveat is that our own universe is not an anti–de Sitter space. With the accelerating expansion of spacetime and the consequent dilution of matter (both light and dark) that this implies, our universe will come in time to resemble a de Sitter, not an anti–de Sitter, universe.

The black hole war might have been won, but Hawking did not immediately concede defeat. Eventually, on 23 April 2007, he paid out on a bet that he had made with Don Page. In 1980 the two had agreed a bet of $1 to £1 that, in essence, black holes destroy information. Hawking had expressed this mathematically through a structure he called the '$-matrix'. Now he declared: 'I concede in light of the weakness of the $.'[14]

The reality check

Battles of wits between rival theorists will always make entertaining reading. In this case, there was an important physical principle at stake, and there can be no doubt that the black hole war helped to promote an important theoretical advance in the form of the holographic principle.

Much of this chapter has been concerned with quantum information and black hole physics, with a consistent thread running through all the key episodes – information and entropy, black holes and the second law of thermodynamics, Hawking radiation, the black hole information paradox, black hole complementarity and the holographic principle, to the triumph of the CFT/AdS duality.

We must now ask the awkward question: how can we tell if any of this theory is right?

By definition, a black hole doesn't give anything away. But we can infer its presence indirectly by the effect it has on visible matter in its neighbourhood. Consequently, there is some astronomical evidence for the existence of black holes.

The accretion of a large quantity of in-falling gas causes the temperature of the gas to rise so high that the material surrounding the black hole emits energetic X-rays which can be detected. This is not to be confused with Hawking radiation, which is emitted by the black hole itself, independently of any in-falling material. Similarly, binary systems composed of a black hole and a companion star can emit X-rays as the stellar material is sucked towards the black hole's event horizon. A strong X-ray source in the constellation Cygnus – called Cygnus X-1 – was one of the first such black hole candidates, with an event horizon thought to be just 26 kilometres in diameter.

So-called 'active galaxies' have cores that are far more luminous over at least some parts of the electromagnetic spectrum. Such activity is believed to result from the accretion of material by a supermassive black hole at the centre of the galaxy. The general consensus is that such supermassive black holes exist at the centres of most galaxies, including our own Milky Way.

But whilst this indirect evidence certainty favours the existence of black holes, there is no astronomical data to tell us what they might actually be like. Consequently, we're not in a position to verify the 'no hair' theorem. We can't measure the temperature of a black hole, or its entropy. We can't determine if black holes really do emit Hawking radiation (all black holes of a size likely to be indirectly detectable will in any case absorb more CMB radiation than they emit). We can't tell if black holes really do evaporate, although the Fermi Gamma-ray Space Telescope is searching for telltale gamma-ray bursts from evaporating primordial black holes.

In *The Hidden Reality*, Greene acknowledges this 'fine print':

We think the answer to each of these questions [concerning information and black hole physics] is yes because of the coherent, consistent, and carefully constructed theoretical edifice into which the conclusions perfectly fit. But since none of these ideas has been subject to the experimenter's scalpel, it is certainly possible (though in my view highly unlikely) that future advances will convince us that one or more of these essential intermediate steps are wrong.[15]

The theorists are no doubt in the best position to judge these things. Few seriously doubt that Hawking radiation is a 'real' phenomenon. But it's worth bearing in mind that a proper description of the quantum nature of black holes really demands a proper quantum theory of gravity. Hawking made great strides by applying general relativity to describe a black hole and then separately applying quantum field theory in the vicinity of the black hole's event horizon. But who is to say what the application of a consistent theory of quantum gravity would yield?

And, just to be clear, there are plenty of examples from the history of science where few theorists have doubted something that has ultimately proved to be quite wrong.

What, then, of the role of quantum information in our description of the universe? What does 'information is physical' really mean? I believe there are two ways in which we can interpret this statement. One is scientific, the second is metaphysical.

The scientific interpretation acknowledges that information is not much different from other physical quantities. But, as such, it is a *secondary quality*. It relies on the properties of physical objects, such as photons with different polarization states or electrons with different spin orientations. In this sense it is like heat or temperature, which is a secondary quality determined by the motions of physical objects. 'Information is physical' means that information must be embodied in a physical system of some kind and processing information therefore has physical consequences. Take the physical system away, and there can be no information.

I obviously have no issue with this.

The metaphysical interpretation suggests that information exists independently of the physical system, that it is a *primary* quality, the ultimate manifestation of an independent reality. 'Information is physical' then acknowledges that in our empirical reality of observation and measurement, information becomes dressed in a clothing of physical properties. This is a bit like suggesting that heat or temperature are the ultimately reality, existing independently but projected into our empirical world of experience in terms of the motions of physical objects.

I have no real issue with this either, so long as we don't pretend that it is science.

Coherence vs correspondence

I have another motive for including the tale of the black hole war in this book. Clearly, this is an area of theoretical physics to which we struggle to apply either the Testability or the Veracity Principle. As there is simply no observational or experimental data to which we can refer the theoretical predictions, we might conclude that this is a branch of theoretical physics that is necessarily speculative or metaphysical in nature.

Clearly, the notion that the entire universe is a hologram projected from information encoded on its boundary belongs firmly in the bucket labelled 'fairy-tale physics'.

But there is a very strong sense that some considerable scientific progress has been made here. Debates have raged over real issues of principle, and these debates have been resolved, it would appear by arriving at the 'right' answer. It really does seem that a singular truth has been established. This is not something that can be lightly dismissed.

How come? If none of this theory is testable, and can't be compared with observational or experimental facts, how has resolving the black hole information paradox helped to establish the 'truth'?

The answer is, I think, reasonably obvious after a moment's reflection. The scientific method is premised on a *correspondence* theory of truth. This is what the Veracity Principle is all about. A scientific statement is held to be true if, and only if, it corresponds to facts that we can establish about empirical reality. This implies that there can be no scientific truth – no scientific right or wrong – without reference to facts about the external world.

But truth, like reality, is a movable feast. There is another kind of truth. We can establish the internal logical consistency of a collection of sentences or propositions – or mathematical structures – independently of the facts. In this *coherence* theory of truth, we seek to establish right and wrong in relation to the theoretical system itself, irrespective of whether or not the theory can be tested by reference to the facts, and so irrespective of whether or not the theory itself is scientifically right or wrong.

Now, of course this is not inconsistent with correspondence to the facts. We would rightly expect that scientific theories of empirical reality should be internally consistent and coherent. We might therefore conclude that a properly scientific theory must be both internally consistent and coherent *and* must make predictions that can be shown to correspond to the facts. But I think you can see how it is possible for theories to be developed which can be demonstrated to be internally consistent and coherent but which *do not* make predictions that can be tested.

The simple truth is that the holographic principle was 'proved' not by reference to the observed behaviours of bits of quantum information and the observed physical properties of black holes, but by reference to a superstring duality; in other words, by reference to another theoretical structure that itself makes no testable predictions.

The truth established in the black hole war is a 'coherence truth', not a 'correspondence truth', and there's a big difference. If we stick rigidly to the definition of science that I outlined in Chapter 1 (and which may have seemed so very reasonable back then), then we must conclude that 'coherence truth' is in itself insufficient to qualify as scientific truth. The history of science is littered with examples of theories that have been demonstrated to be internally consistent and coherently true (and which few have doubted), but which have been shown to be ultimately worthless as scientific theories.

Now, I am by no means suggesting here that the holographic principle should be considered worthless. There may well be applications of the principle in information theory that can be tested as techniques in quantum computing are developed.

I guess all I'm really asking for is the exercise of a little scientific scepticism. When we read popular science books and articles and watch television documentaries that tell us that information is the ultimate reality and that the universe is a hologram, let's just put all this into a proper perspective.

Let's stay focused on the nature of the 'truth' that is being communicated.

11

Ego Sum Ergo Est

I Am Therefore It Is: the Anthropic Cosmological Principle

A theorist goes astray in two ways: 1. The devil leads him by the nose with a false hypothesis. (For this he deserves our pity.) 2. His arguments are erroneous and sloppy. (For this he deserves a beating.)

Albert Einstein[1]

In the last four chapters we have had an opportunity to review the various approaches that theorists have taken in their attempts to resolve some of the more stubborn problems with the current authorized version of reality.

We have seen how attempts to resolve the quantum measurement problem have led to many worlds. We have seen how attempts to push beyond the standard model of particle physics and find rhyme or reason for its twenty experimentally determined parameters have led to SUSY and superstring/M-theory. We have seen how SUSY and various braneworld scenarios in M-theory suggest solutions for the hierarchy problem. In search of potential dark matter candidates we examined the lightest supersymmetric particles predicted by SUSY and the lightest Kaluza–Klein particles, thought to be projected from M-theory's hidden dimensions. We saw how the multiverse resolves the problem of dark energy and the cosmological constant. In the cosmic landscape, ours is but an ordinary, if not rather mundane, universe among a vast multiplicity of universes.

Despite their speculative or metaphysical nature, in the context in which I've presented them so far these contemporary theories of physics still conform to the Copernican Principle. They do not assume an especially privileged role for us as human observers.

Perhaps you'll then be surprised to learn that one approach to resolving the fine-tuning problem, an approach growing in importance and gaining support within the theoretical physics community, puts human beings firmly back into the equation. It is called the anthropic cosmological principle, where 'anthropic' means 'pertaining to mankind or humans'.

I should say upfront that this is all rather controversial stuff. Many scientists have argued that the anthropic cosmological principle is neither anthropic nor a principle. Others have argued that it is either dangerous metaphysics or completely empty of insight. Either a false hypothesis, deserving of Einstein's pity, or erroneous and sloppy thinking, deserving of a beating.

I believe that the anthropic cosmological principle is symptomatic of the malaise that has overtaken contemporary theoretical physics. We'd better take a closer look.

The carbon coincidence

In his *Discourse on Method*, first published in 1637, the French classical modern philosopher René Descartes tried to establish an approach to acquiring knowledge of the world by first eschewing all the information delivered to his mind by his senses. He had decided that his senses couldn't be trusted. He cast around looking to fix on something about which he could be certain.

After some reflection, he decided that that something was his own mind. And, he reasoned, given that he possesses a mind, then in some form or another he must exist. *Cogito ergo sum*, he declared: I think therefore I am.

Human beings are carbon-based life forms that have evolved certain mental capacities. We are conscious and self-aware, and, like Descartes, we are able to reflect intelligently on the nature of the physical universe we find around us. It seems a statement of the blindingly obvious that, whatever it is and wherever it comes from, the physical universe supports the possibility that we could (and, indeed, do) exist. We exist therefore the universe must be just so. We might adapt Descartes' famous saying thus: *ego sum ergo est* – I am therefore it is.

The problem is that as soon as we put human beings (or, at the very least, the possibility of cognitive biological entities) back into the equation in this way, we acquire a perspective that makes the universe look like an extraordinary conspiracy.

One of the most notable examples of a 'coincidence' of the kind that betrays conspiracy in the physical mechanics of the universe was identified by the physicist Fred Hoyle in the early 1950s. It concerns the process by which carbon nuclei are produced in the interiors of stars.

The primordial big bang universe (or the steady-state universe favoured at the time by Hoyle) contains only hydrogen and helium and trace amounts of slightly heavier elements such as lithium. In the early 1950s, the relative abundances of heavier elements were therefore something of a mystery. How are these elements formed?

Hoyle supposed that at the high temperatures and pressures that prevail in the centres of stars, the primordial hydrogen and helium would get further 'cooked'. These light nuclei would fuse together in a series of reactions to form successively heavier nuclei, in a process now called *stellar nucleosynthesis*.

When two hydrogen nuclei fuse together to form a helium nucleus, two of the four protons transform into neutrons, and energy is released. This energy holds the star up against further gravitational collapse, and the star settles down into a period of relative stability.

As the supply of hydrogen becomes depleted, however, the energy released from such fusion reactions is no longer sufficient to resist the force of gravity. A star with enough mass will blow off its outer layers and its core will shrink. The temperature and pressure in the core will rise, eventually triggering fusion reactions involving helium nuclei.

But at this point we hit a snag. Fusing hydrogen nuclei (one proton) and helium nuclei (two protons and two neutrons) together to make lithium is not energetically possible. A lithium nucleus with three protons and two neutrons is unstable: it needs one or two more neutrons to stabilize it. Fusing two helium nuclei together to make beryllium is similarly impossible – a beryllium nucleus with four protons and four neutrons is likewise unstable. It needs another neutron.

In a state of rising panic, we skip over lithium and beryllium and look to the next element in the periodic table. What about carbon? A carbon nucleus has six protons and six neutrons. This would seem to require fusing together three helium nuclei. This is energetically possible, but the chances of getting three helium nuclei to come together in a simultaneous 'three-body' collision are extremely remote. It's much more feasible to suppose that two helium nuclei first fuse to form an unstable beryllium nucleus, which then in turn fuses with another helium nucleus before it can fall apart. This sounds plausible on energy grounds but the odds don't look good. The beryllium nucleus tends to fall apart rather too quickly.

Yet here we are, intelligent beings evolved from a rich carbon-based biochemistry. Given that we exist, carbon must somehow be formed in higher abundance, despite the seemingly poor odds.

Hoyle reasoned that the odds *must* somehow get tipped in favour of carbon formation. He therefore suggested that the carbon nucleus must possess an energetic state that helps greatly to enhance the rate of the reaction between the unstable beryllium nucleus and another helium nucleus, thereby producing carbon faster than the beryllium nucleus can disintegrate. Such an energetic state is called a 'resonance'. Hoyle estimated that the carbon nucleus must have a resonance at an energy of around 7.7 MeV. It was subsequently discovered at 7.68 MeV. The reaction is called the triple-alpha process.*

This struck Hoyle as remarkable. If the carbon resonance were slightly higher or lower in energy, then carbon would not be formed in sufficient abundance in the interiors of stars. There would therefore be insufficient carbon in the debris flung from those stars that are ultimately destined to explode in spectacular supernovae. The second-generation star systems that formed from this debris would then hold planets with insufficient carbon to allow intelligent, carbon-based life forms to evolve.

Change the energy of the carbon resonance by the slightest amount, and we could not exist. Hoyle wrote:

> Would you not say to yourself, 'Some super-calculating intellect must have designed the properties of the carbon atom, otherwise the chance of my finding such an atom through the blind forces of

---

* This is because the helium nucleus, consisting of two protons and two neutrons, is also an alpha particle. The process involves three alpha particles combining together to form a carbon nucleus.

nature would be utterly minuscule.' Of course you would ... A common sense interpretation of the facts suggests that a superintellect has monkeyed with physics, as well as with chemistry and biology, and that there are no blind forces worth speaking about in nature. The numbers one calculates from the facts seem to me so overwhelming as to put this conclusion almost beyond question.[2]

The Goldilocks enigma

The carbon coincidence is just the beginning. It occurs because of a delicate balance between the strength of the strong force and the energetics of nuclear reactions involving protons. In *Just Six Numbers*, British astrophysicist Martin Rees identified a series of six dimensionless physical constants and combinations of constants that determine the nature and structure of the universe we inhabit. Change any one of these numbers by just 1 per cent and, Rees argued, the universe that resulted would be inhospitable to life. If the constants were not so fine-tuned, we could not exist to observe the universe and ponder on its remarkable cosmic coincidences.

These six numbers include ε, the fraction of the mass of the four protons that is released as energy when these fuse together to form a helium nucleus inside a star. ε determines the amount of energy released by a star like our own sun and the subsequent chain of nuclear reactions responsible for the production of other chemical elements. Like the carbon coincidence, it depends on the strength of the strong force. If too little energy is released, the planetary system orbiting the star remains cold and lifeless. Too much, and the planetary system is hot and lifeless.

The set of numbers also includes N, the ratio of the strength of the electromagnetic force to the strength of the force of gravity. This is a large number (10^{36}), and says that the mechanics of the atom are dominated by electromagnetic forces – gravity is irrelevant. But gravity is cumulative. Gather lots of atoms together and it adds up. N determines the relationship between the behaviour of matter at atomic and subatomic levels and matter at the levels of planets, stars, galaxies and clusters of galaxies. Make the force of gravity just a little bit larger in relation to the force of electromagnetism and the universe would be smaller and would evolve much faster. There would be no time for biology to develop. Make gravity a little weaker and there would be no stars.

The density parameter, Ω, is the ratio of the density of mass-energy to the critical value required of a flat universe. It depends on the balance between gravity and the rate of expansion of the universe. Too much mass-energy and the result is a closed universe which expands a little but then contracts rather quickly: too quickly for life to gain a foothold. Too little mass-energy and the result is an open universe which expands too quickly to support the evolution of galaxies.

Likewise, the cosmological constant, Λ, seems similarly fine-tuned for life. Although it was something of a shock to cosmologists in the late 1990s to discover that Λ isn't precisely zero and that the expansion of the universe is accelerating, the value of the constant is still extremely small. Ridiculously small according to quantum theory. But if it was any larger, the universe would be open and there would be no stars, no galaxies and no life.

Of course, we can trace the large-scale structure of the visible universe – galaxies and clusters of galaxies – right back to the quantum fluctuations that prevailed during the inflationary epoch. These ripples are slight – about one part in 100,000. This variation is captured in the ratio $\mathcal{Q}$, derived from the energy required to break up large galactic clusters or superclusters and the energy of the rest mass of such structures. If $\mathcal{Q}$ were smaller, there would be no large-scale structures. Make it larger and the universe would consist only of supermassive black holes.

Rees' sixth number is the simplest of the set. It is $\mathcal{D}$, the number of spatial dimensions (not including 'hidden' dimensions demanded by superstring/M-theory). There are no one- or two-dimensional complex biological entities, simply because complex biology demands a minimum of three dimensions. And a value of $\mathcal{D}$ of 3 is the only number compatible with the inverse-square laws of gravity and electromagnetism.

The paranoia runs deep. Virtually everywhere we turn we're faced with an apparently phenomenal fine-tuning of the universe's physical constants and laws. If the weak force were stronger or weaker, then the primordial abundances of hydrogen and helium would have been very different, and subsequent stellar nucleosynthesis would not have produced the ingredients needed to sustain life. If the mass of the neutron wasn't slightly larger than the proton … And so on and on.

In the fairy tale, Goldilocks finds that baby bear's porridge is just the right temperature, that baby bear's chair is just the right size and that baby bear's bed has just the right softness. And what happens when we put human beings (or, at the very least, the possibility of biology) back into the picture? We find that the universe is not 'just right', it appears extraordinarily fine-tuned for life.

This is the Goldilocks enigma.

The weak anthropic principle

Although there have been many examples of 'anthropic' reasoning throughout history, the notion of an anthropic principle was first introduced by the Australian theorist Brandon Carter. Whilst studying for his PhD at Cambridge University in 1967, Carter had become absorbed by the challenge of understanding the origin of the numerical coincidences that seem to dominate cosmology and physics. Influenced by Wheeler at Princeton, he circulated lecture notes on the subject informally among colleagues, eventually publishing his ideas in 1974.

Carter intended the anthropic principle as a direct challenge to the Copernican Principle. He presented his arguments at an International Astronomical Union symposium in Cracow on 10–12 September 1973. The conference had been dedicated to commemorate the five hundredth anniversary of Copernicus' birth.

Whereas the Copernican Principle insists that intelligent life occupies no privileged position in the cosmos, the anthropic principle suggests that our status as conscious observers of our universe *demands* at least some form of privileged perspective.

Carter offered two versions. The first is the *weak anthropic principle*: 'we must be prepared to take account of the fact that our location in the universe is necessarily privileged to the extent of being compatible with our existence as observers'.[3]

This seems like common sense, if a bit of a tautology. In essence, it is a statement that relates to the notion of *observer self-selection*. The universe we observe must, by definition, include observers like us capable of observing it. Our observations are necessarily biased by virtue of our own existence. We are (again, by definition) unable to observe a universe in which observers like us cannot exist.

In Carter's definition, the word 'privileged' is used in a relatively mild sense. This is entirely compatible with the notion that intelligent observers are a perfectly natural phenomenon. No matter how improbably the physical universe appears to be fine-tuned and no matter how unlikely the facts of biological evolution through natural selection, the bottom line is that we're here.

Now, the key question is what – if anything – the weak anthropic principle has to say about the *reason* we're here. One argument is that we're here because, by happy accident or the operation of some complex natural physical mechanisms we have yet to fathom, the parameters of the universe just happen to be compatible with our existence.

In this argument, intelligent life remains a *consequence* of the nature and structure of our universe (it is therefore I am). But it leaves us in the rather unsatisfactory position of having no real explanation under present understanding as to precisely *why* the universe is the way it is.

Maybe you can already sense where this is leading. If we have no explanation for the fine-tuning of the universe, perhaps this is because the universe isn't really fine-tuned after all. What if the universe we inhabit is but one of an infinite or near-infinite number of parallel universes in which all manner of different combinations of physical parameters are possible?

In most of these universes the parameters are incompatible with the existence of intelligent life. The operation of selection bias means that, no matter how atypical or improbable the parameters of the universe we find ourselves in, we shouldn't be surprised to find that this is nevertheless precisely the universe we observe.

The anthropic multiverse

Carter used the weak anthropic principle to argue for the existence of what he called an *ensemble* of universes – meaning simply that there are many (maybe an infinite number) – without specifying precisely what these were or where they might have come from. There are many different possibilities, some of which were discussed in Chapter 9. Readers interested in a more comprehensive review of the different multiverse theories should consult Brian Greene's *The Hidden Reality*.

Of course, this is the reason why the anthropic principle has become so popular among theorists in recent years. It shouldn't come as any real surprise that those theorists who favour the cosmic landscape of superstring/M-theory's 10^{500} different ways of constructing a universe have rushed to embrace anthropic reasoning.

What I find quite remarkable is that the observer selection bias summarized in the anthropic principle is sometimes used as a kind of *justification* for landscape theories, as though it were an important piece of observational evidence in support of them! The logic runs: the fact that our universe seems highly improbable *must* mean that observer selection bias is operating in a multiverse of possibilities.

The relationship is mutual. The landscape is also used to lend credibility to the anthropic principle, as Susskind claims: 'Whether we like it or not, this is the kind of behavior that gives credence to the anthropic principle.'[4]

Although this marriage between the landscape and anthropic logic might initially have been rather forced, it does seem to be a marriage made in heaven. In *The Cosmic Landscape*, Susskind writes:

> Until very recently, the anthropic principle was considered by almost all physicists to be unscientific, religious, and generally a goofy misguided idea. According to physicists it was a creation of inebriated cosmologists, drunk on their own mystical ideas ... But a stunning reversal of fortune has put string theorists in an embarrassing position: their own cherished theory is pushing them right into the waiting arms of the enemy ... The result of the reversal is that many string theorists have switched sides.[5]

The strong anthropic principle

Carter's second version of the anthropic principle is the *strong anthropic principle*: 'the Universe (and hence the fundamental parameters on which it depends) must be such as to admit the creation of observers within it at some stage'.[6]

In this definition the key word is 'must', and one (somewhat irresistible) interpretation restores intelligent life to its pre-Copernican, fully privileged status. The strong anthropic principle tips us towards teleology – philosophical positions that perceive ultimate purposes,

JIM BAGGOTT

final causes and deliberate design in nature. It puts 'us' firmly back at the centre of the picture.

To be fair to Carter, both his weak and strong anthropic principles were based on the notion of intelligent observers, not necessarily carbon-based life forms like ourselves. He later regretted his use of the word 'anthropic'.

But the cat was out of the bag. When theorists John Barrow and Frank Tipler popularized the principle in their book *The Anthropic Cosmological Principle*, first published in 1986, they made no bones about making both the weak and strong versions all about carbon-based life forms. This was a carefully constructed scholarly survey of the literature on the subject, and although Barrow and Tipler did not argue for any particular position, they extended the principle into some highly speculative, extreme versions and helped to muddy the waters considerably.

The argument from design

The strong form of the anthropic principle drags us into some highly contentious and emotionally charged territory.

There's a good chance that, having made it this far, you're a reader who happens to be interested in what contemporary science is saying about the nature of the physical world around us. You've hopefully taken the trouble to absorb the six Principles presented in Chapter 1, and whether you agree with them or not, I would assume you're reasonably clear in your own mind about what science is, broadly speaking.

I'm hoping that we can therefore agree that intelligent design is not science.

Intelligent design is a variation of 'creation science' (a non-sequitur if ever there was one). As an idea it has a relatively long history, but its modern form was developed by a group of American creationists and is today most closely associated with a non-profit organization called the Discovery Institute, founded in 1990 and based in Seattle, Washington. The institute promotes a number of projects across a range of areas. In the field of science and culture, its website declares its agenda as follows:

Scientific research and experimentation have produced staggering advances in our knowledge about the natural world, but they have also led to increasing abuse of science as the so-called 'new atheists' have enlisted science to promote a materialistic worldview, to deny human freedom and dignity and to smother free inquiry. Our Center for Science and Culture works to defend free inquiry. It also seeks to counter the materialistic interpretation of science by demonstrating that life and the universe are the products of intelligent design and by challenging the materialistic conception of a self-existent, self-organizing universe and the Darwinian view that life developed through a blind and purposeless process.[7]

Okay, so we're clear about that.

The Discovery Institute's ambition to get intelligent design taught in science classes alongside Darwin's theory of evolution by natural selection led in 2005 to the landmark case of Tammy Kitzmiller et al. vs Dover Area School District et al., which was heard in a US district court in Pennsylvania. Conservative Republican Judge John E. Jones III ruled that intelligent design is *not* science. A Dover school board decision to oblige its staff to teach intelligent design was ruled to be unconstitutional. In a subsequent school board election, the eight board members who had voted for this decision were ousted.

When I was young, I thoroughly enjoyed the 1960 film *Inherit the Wind*, starring Spencer Tracy and Frederic March as protagonists locked in a courtroom debate about the teaching of evolution in an American school. The film was based on the 1955 play by Jerome Lawrence and Robert Edwin Lee, which was itself an accurate fictional account of the 1925 trial of John Thomas Scopes in Dayton, Tennessee. Scopes stood 'accused' of teaching the theory of evolution. He was found guilty and fined $100, but the verdict was subsequently overturned following an appeal to the Supreme Court.

I had imagined that these kinds of courtroom battles were things of history. When I first heard about the Dover School case, eighty years after the Scopes trial, I was utterly shocked. Yes, these were two very different cases. In the first, it was Darwin in the dock. In the second, it was intelligent design. But at stake in both cases was the notion that religious or theistic beliefs should *not* inform efforts to understand the 'truths' of our physical, chemical and biological world. I had assumed that

271

such a notion was by now self-evident in civilized societies and that there had long since ceased to be a case to answer, either way. I was wrong.

Following the 2005 ruling, the Discovery Institute has had to adopt a more subtle strategy. It no longer seeks to require that intelligent design be taught in science class. Instead it attempts to ensure that evolution is presented as a scientific theory that can (and no doubt should) be critically examined and challenged. This is laudable, and very much part of the scientific process, although it would be unusual for high school students to be taught to question and challenge *everything* they were told in class. Science progresses because we are able to 'bank' some scientific truths – we can accept them as valid and move on. If we stopped to question everything we think is scientifically true, then we wouldn't make much progress.

But the institute also suggests that there is nothing unconstitutional in teachers 'discussing' (rather than formally teaching) the 'scientific theory of design'. It encourages such discussion, under the guise of academic freedom. So, the only way to prevent an insidious creeping of creationism into school science discussions is, once again, to be clear that intelligent design is not science. It belongs instead in a philosophy or theology class.

And this is why I find the strong anthropic principle deeply worrying. By its very nature, it is intended as a counterpoint to the Copernican Principle. It therefore undermines the very basis of science as it has been practised over the last five hundred years. It is surely a gift to those who, like the members of the Discovery Institute, seek to promote discussion of intelligent design as a valid 'scientific' theory.

The science historian Helge Kragh voiced similar concerns almost a quarter of a century ago. In his review of Barrow and Tipler's *The Anthropic Cosmological Principle* he commented:

> Under cover of the authority of science and hundreds of references Barrow and Tipler, in parts of their work, contribute to a questionable, though fashionable mysticism of the social and spiritual consequences of modern science. This kind of escapist physics, also cultivated by authors like Wheeler, Sagan and Dyson, appeals to the religious instinct of man in a scientific age. Whatever its merits it should not be accepted uncritically or because of the scientific brilliancy of its proponents.[8]

Writing in 2011, Kragh noted:

> I was convinced that within a decade physicists would lose interest in anthropic considerations and leave them to where they properly belonged, namely to philosophers and theologians. I was seriously wrong.[9]

The John Templeton Foundation

The feelings of discomfort generated by the strong anthropic principle remind us that modern theoretical physics has drifted far from its notional purpose: to provide an interpretation of the physical world that we can understand and accept as scientifically true.

To a certain extent, physical theorists have always tended to sail close to the wind and wax rather speculatively, philosophically or even theologically at times. It comes with the territory. The subject is after all concerned with the 'big questions', and in the absence of hard scientific evidence, there's nothing in principle wrong with a few personal opinions.

But the relationship between science and religion remains uneasy and is sometimes antagonistic, as evidenced by the Dover School case. Most scientists accept that there are fundamental questions of purpose, existence and meaning that science cannot pretend to provide answers for. These are questions that should be dismissed as irrelevant non-questions, or else the answers should be sought from some system of belief that is not scientifically based.

Despite this unease, many contemporary theorists have comfortably set up camp in the space where science meets philosophy meets theology. Some of these theorists have written popular books, using the 'authority of science', as Kragh puts it, to advance their own personal world views. Many of these are great books (I have them on my shelves), and reading them as a graduate and subsequently postgraduate student inspired me eventually to try my own hand at writing about science.

But it is important to be aware of these authors' likely agendas. I don't mean to suggest that these agendas might be in some way hidden, or that there's some grand conspiracy in play. What I mean is that when we read the next best-selling popular science book, we might be interested in understanding roughly where the author has set up camp.

The John Templeton Foundation was established by Sir John Templeton in 1987, with an endowment valued in 2011 at $2.3 billion. Templeton, an American-born British investor and philanthropist, died in 2008, aged 95. The foundation is overtly religious or theological in nature, but through the award of annual prizes and grants, it supports science and invests in research on the 'big questions'. Templeton's philanthropic vision includes the following observations:

> There may be significant promise in supporting a wide range of careful and rigorous research projects by well-regarded scientists on basic areas with theological relevance and potential ... to examine or foster the idea that through an expanded search for more knowledge, in which we are open-minded and willing to experiment, theology may produce positive results even more amazing than the discoveries of scientists that have electrified the world ... in the 20th century.[10]

The Templeton Prize is now valued at £1 million.* Recipients have included religious figures (the Dalai Lama was awarded the prize in 2012), but also many physicists – Stanley Jaki (1987), Carl Friedrich von Weizsäcker (1989), Paul Davies (1995), Freeman Dyson (2000), John Polkinghorne (2002), George Ellis (2004), Charles Townes (2005), John Barrow (2006), Bernard d'Espagnat (2009) and Martin Rees (2011).

Some of these Templeton prize laureates either work at the intersection of physics and theology or have feet in both camps, as it were. Jaki was an astrophysicist and Benedictine priest (he died in

---

* If this seems rather generous (the Nobel Prize is valued at about £750,000, which is shared between recipients), then we should probably stop to reflect on the new Fundamental Physics Prize, established on 31 July 2012 by Russian entrepreneur Yuri Milner. Milner's new Foundation is 'dedicated to advancing our knowledge of the Universe at the deepest level by awarding annual prizes for scientific breakthroughs, as well as communicating the excitement of fundamental physics to the public.' Prizes are valued at $3 million (almost £2 million), and many notable string theorists were among the nine to receive its first awards. In December 2012, Stephen Hawking was awarded a special prize of $3 million and a second $3 million prize was shared by a select group of experimental physicists working at CERN's LHC. For more information, see www.fundamentalphysicsprize.org.

2009). Polkinghorne is a mathematical physicist and Anglican priest. Ellis is a South African cosmologist and active Quaker, and in the 1970s and 1980s was a vigorous opponent of apartheid.

It should come as no surprise that the Templeton Foundation would seek to recognize and reward those scientists who have reconciled their science with their religious beliefs and who have contributed to the development of what Templeton himself referred to as 'new spiritual information'. But what about the others?

Both Davies and Barrow are very successful popular science writers. In books such as *Other Worlds*, *God and the New Physics* and *The Goldilocks Enigma*, Davies has made no secret of his particular world view. Thus, in *The Goldilocks Enigma*, he writes:

> Many scientists will criticize my … inclinations towards [theories in which the universe is purposefully driven to develop life and mind and in which life and mind somehow 'create' the universe in a closed causal loop] as being crypto-religious. The fact that I take the human mind and our extraordinary ability to understand the world through science and mathematics as a fact of fundamental significance betrays, they will claim, a nostalgia for a theistic world view in which humankind occupies a special place. And this even though I do *not* believe *Homo sapiens* to be more than an accidental by-product of haphazard natural processes. Yet I do believe that life and mind are etched deeply into the fabric of the cosmos, perhaps through a shadowy, half-glimpsed life principle, and if I am honest I have to concede that this starting point is something I feel more in my heart than in my head. So maybe that is religious conviction of sorts.[11]

Barrow was awarded the Templeton Prize in recognition of his contributions to debates on spirituality and the purpose of life. The 'chronicle' issued by the foundation at the time of his award makes special mention of his role as co-author of *The Anthropic Cosmological Principle*:

> This book has been enormously influential in discussions between religious and scientific perspectives on the universe. It has been cited very heavily across the spectrum of scholarly study from

studies of natural theology, philosophy, physics, mathematics, and astronomy. Of particular interest to the theology–science interface is the detailed history of design arguments and natural theology, to which Dr Barrow contributed with the modern cosmological forms of the anthropic principle.[12]

The award of the 2011 prize to Martin Rees caused something of a stir. Rees declares that he has no religious beliefs at all, but maintains an appreciation of the culture of the Anglican church. He is an irregular churchgoer and, in his own words, is not 'allergic to religion'. Rees has had a highly illustrious career, but appears to have made no overt contribution to 'new spiritual information' in areas of theological relevance. But this is not quite how the Templeton Foundation saw it. According to the press release issued at the time of the award announcement, 'By peering into the farthest reaches of the galaxies, Martin Rees has opened a window on our very humanity, inviting everyone to wrestle with the most fundamental questions of our nature and existence.'[13]

Now, it is quite possible to peer into the farthest reaches of the galaxies without making any pronouncements on fundamental questions of our nature and existence, and many similarly illustrious astrophysicists have done just this. It is no doubt Rees's work on 'cosmic coincidences' and his leaning towards anthropic reasoning that made his candidacy appear attractive to the foundation. In his recommendation, Robert Williams, president of the International Astronomical Union, noted that Rees 'is very unusual in that he constantly touches on spiritual themes without dealing explicitly with religion'.[14]

So, what's my point? I want to be clear that I have no real problem with the activities of the Templeton Foundation. It appears to be a lot less insidious than the Discovery Institute (although not all commentators agree) and it funds some excellent work in theoretical physics and philosophy. Rees himself declared: 'They are very nice people who are doing things which are within their agenda.'[15]

There are of course issues with how acceptance of such an award by an eminent scientist is perceived – Rees is a former president of the Royal Society and Astronomer Royal. Arch-atheist Richard Dawkins, no friend of the Templeton Foundation, pointed out that the award would look a lot better on Templeton's CV than it would on Rees's.[*]

Genuine altruism is very hard to find, and all funding bodies seek to drive an agenda of some kind. The Templeton Foundation certainly doesn't attempt to influence the nature and outcomes of the research that it funds. But by actively promoting scientific and philosophical debate on subjects of 'spiritual' or theological relevance, it does raise the profile of these subjects and gives them a prominence that they may not necessarily deserve.[**]

My concern is only that we should always try to be aware of what we're dealing with. Theoretical physics is in part concerned with the 'big questions'. This is what makes it fascinating. But most theoretical physicists are complete human beings – they are interested in things and believe things that often draw them beyond the boundaries of even a loosely defined scientific method. They want to talk and write about these things, to contribute to a broader dialogue relevant to our human concerns, in ways that some – like the Templeton Foundation – perceive as 'spiritual'.

In this way theorists add to our cultural richness. But these are *very* murky waters, and we should wade into them with eyes wide open. The line between use and abuse of the 'authority of science' to promote a particular outlook (on life, the universe and everything) is extremely thin.

The reality check

In his 2002 book *Anthropic Bias*, Swedish philosopher Nick Bostrom commented that:

> The 'anthropic principles' are multitudinous – I have counted over thirty in the literature. They can be divided into three categories:

---

[*] The Templeton Foundation acquired the services of leading PR company Bell Pottinger to handle the media buzz that was provoked by the announcement.

[**] For example, the Templeton foundation publishes *Big Questions Online* (www. bigquestionsonline.com), on which can be found a recent (10 July 2012) posting by Steven M. Barr, Professor of Physics at Delaware University, titled: 'Does Quantum Physics Make it Easier to Believe in God?'.

those that express a purported observation selection effect, those that state some speculative empirical hypothesis, and those that are too muddled or ambiguous to make any clear sense at all.[16]

I don't think we need to waste time debating whether the strong anthropic principle, or indeed any similarly structured principle, is scientific. Any structure designed completely to overturn the Copernican Principle and restore some kind of privileged status to intelligent observers (whether human or not) goes against the grain of nearly five hundred years of scientific practice.

Now, the success of past practice certainly doesn't make this practice right, not least because the six Principles described in Chapter 1 leave plenty of room for manoeuvre and interpretation. But on the basis of the interpretation I have given and used throughout this book, I would conclude that the strong anthropic principle is not science.

This leaves us to consider the weak form of the principle. Should we deem this to be scientific? Can the weak anthropic principle be used to make testable predictions? How should we interpret the use of the principle in the context of the multiverse?

Time for a final reality check.

Although there are many claims for predictions made on the basis of anthropic reasoning, there are two examples that stand out. These are the 'carbon coincidence', which led to Hoyle's prediction of a carbon resonance at 7.7 MeV; and a prediction concerning the magnitude of the cosmological constant made by Steven Weinberg in 1987. The former example is outlined above, so let's quickly back-fill on the latter.

To a large extent, we have already examined the rationale that Weinberg used in our earlier discussion of fine-tuning. The effect of a non-zero cosmological constant is to accelerate the expansion of spacetime. If the cosmological constant is too large, spacetime expands too quickly and galaxies and clusters of galaxies cannot form. In 1987, the cosmological constant was assumed to be zero, but the possibility of vacuum fluctuations of the kind predicted by quantum theory suggest something rather larger. A factor 10^{120} larger, in fact.

Weinberg came up with an alternative approach: 'Perhaps Λ must be small enough to allow the universe to evolve to its present nearly empty and flat state, because otherwise there would be no scientists to worry about it.'[17]

Whereas Hoyle's prediction was turned into an example of anthropic reasoning post hoc (Hoyle did not invoke observer self-selection at the time he made his prediction in 1953, although he was subsequently comfortable with this logic), Weinberg's paper was overtly anthropic in nature. It is titled 'Anthropic Bound on the Cosmological Constant', and he referred readers interested in learning more about anthropic reasoning to Barrow and Tipler's book, Paul Davies's 1982 book *The Accidental Universe* and a 1983 paper by Carter.

Weinberg refined his prediction in a paper published two years later. At the time, there was no evidence for a small positive cosmological constant, but he speculated that the anthropic upper bound was large enough to show up eventually in astronomical observations. Nine years later, observations of Type Ia supernovae confirmed that his instincts were correct.

The carbon coincidence and the anthropic upper bound on the cosmological constant appear on the surface to be good examples of anthropic reasoning leading to genuine predictions that were subsequently tested and verified through observation and experiment. But before we rush to embrace the weak anthropic principle as science, let's dig a little deeper.

As I mentioned, Hoyle did not use anthropic reasoning in making his prediction. What, then, did he do? He took a perfectly good scientific observation – the relative abundance of carbon in the universe – and argued that there must be a valid scientific reason for this. If his ideas about stellar nucleosynthesis were right, then there had to be a physical mechanism by which carbon is formed in the interiors of stars. The most logical physical mechanism involves the existence of a resonance at an energy that will enhance the rate of carbon formation over the rate of disintegration of the beryllium nuclei. The fact that we are intelligent, carbon-based life forms is neither here nor there.

A study of the history of Hoyle's prediction led Kragh to conclude:

> Only in the 1980s, after the emergence of the anthropic principle, did it become common to see Hoyle's prediction as anthropically significant. At about the same time mythical accounts of the prediction and its history began to abound. Not only has the anthropic myth no basis in historical fact, it is also doubtful if the excited levels in carbon-12 and other atomic nuclei can be

used as an argument for the predictive power of the anthropic principle.[18]

Similarly, even though Weinberg's logic was overtly anthropic, he could have reached the same conclusions without referring to the anthropic principle, The question again concerns physical mechanism, not the fact of human existence. The mathematical formulae in Weinberg's 1987 paper concern only physical parameters, such as the density of mass–energy in the universe, the gravitational constant, and so on. Weinberg took some perfectly good scientific observations – the relative abundance of galaxies in the universe – and argued that there must be a valid scientific reason for this. The most logical physical mechanism demands a practical upper bound on the cosmological constant.

Writing in 2008, the theorist Lee Smolin observed: 'Just as Hoyle's argument has nothing to do with life, but is only based on the observed fact that carbon is plentiful, Weinberg's first argument has to do only with the observed fact that galaxies are plentiful.'[19]

These are examples of relatively straightforward scientific deduction from observed facts. Yes, I know 'observed' must mean we're here to make the observations. However, whilst it is certainly important to be aware of selection effects in scientific observation (and of course, we run head-on into such effects in quantum theory), we don't tend to be in the habit of putting *every* observation in a human context, even though no observation of any kind is possible *without* humans.

All scientific knowledge is, after all, intrinsically human knowledge.

At issue, then, is not whether the weak anthropic principle is scientific. Rather, it is whether it actually adds any real *value* to scientific reasoning. To answer this question we would need to find an example of a scientific (and therefore testable) prediction that is unambiguously driven by anthropic reasoning. To my mind, this would have to be a prediction that depends on a selection bias stronger than simply observers observing the facts of nature, since it seems we should always be able to lift consideration of selection effects out of the logic without greatly impairing our ability to reach a valid conclusion.

I know of no such anthropic prediction. And, I suspect, given that science as it is widely practised is based on detached observation and experiment, the circularity of the logic means that any such anthropic prediction wouldn't actually be scientific. In other words, I suspect that the Copernican Principle is too deeply embedded in the very fabric and meaning of science that to try to eliminate it, even through the agency of a relatively mild form of privilege, renders the result unscientific.

Science is just not set up this way.

All things being equal

Finally, we must now turn our attention back to the relationship between the anthropic principle and the multiverse, which is the principal reason for the recent surge of interest in anthropic reasoning.

In an interesting twist, it has been argued that the application of the weak anthropic principle to the notion of the multiverse actually 'saves' science from the threat posed by intelligent design. The subtitle of Susskind's *The Cosmic Landscape* reads: 'String Theory and the Illusion of Intelligent Design'. We're encouraged to accept the anthropic multiverse because this is the best solution to the fine-tuning problem. Reject it and we're stuck with intelligent design as the only alternative explanation for the universe we inhabit.

In *Anthropic Bias*, Bostrom argues that, all things being equal, a multiverse theory is indeed more probable:

> ... consider a single-universe theory h_U on which our universe is fine-tuned, so that conditional on h_U there was only a very small probability that an observer-containing universe should exist. If we compare h_U with a multiverse theory h_M, on which it was quite likely that an observer-containing universe should exist, we find that if h_U and h_M had similar prior probabilities, then there are prima facie grounds for thinking h_M to be more probable than h_U given the evidence we have.[20]

Bostrom uses Bayesian decision theory to make his point. Similar arguments have been made by superstring theorists, so it's worth a short excursion to explain this logic.

Thomas Bayes was an eighteenth-century mathematician and Presbyterian minister who developed a theorem concerning probability, published after his death in 1761. In its modern form, Bayes's theory is sometimes used to evaluate and compare scientific theories in relation to observational and experimental evidence. A scientific theory is allocated a prior probability, which we can think of as expressing a scientist's 'degree of belief' in it. After reflecting on some relevant observational or experimental data, we compute a posterior probability. If this is greater than the prior probability then the evidence is clearly in favour of the theory and our degree of belief in it increases.

If we are confronted by two distinct theories with equal prior probabilities, and the evidence increases the posterior probability of one of them above the other, then we clearly have a means to choose between them. We keep the theory with the higher posterior probability and reject the other (at least, until we get more evidence).

The point that Bostrom is making is that if our two theoretical models under consideration (which he calls h_U and h_M) both have equal prior probabilities, then simple statistical probability favours the multiverse model. The fine-tuned single universe h_U is simply much less probable than an observer-selected universe existing among a great multiplicity of possibilities, h_M.

I have no argument with this, as it stands. But if I take h_U to be the current authorized version of reality – in the form of the standard models of particle physics and big bang cosmology – and h_M to be any contemporary multiverse theory, then I have a *big* problem. These two theories do *not* have equal prior probabilities. By virtue of all the observational and experimental facts that support it, my degree of belief in h_U is high.

In case we've forgotten, let's quickly remind ourselves of the status of the multiverse described by superstring/M-theory. We first assume that elementary particles can be represented as vibrations in filaments of energy. We assume a supersymmetric relationship between fermions and bosons. We assume that superstring theory's extra spatial dimensions are compactified in a Calabi–Yau space. We accept the M-theory conjecture. We assume that our universe is but one of a large number (possibly an infinite number) of inflating spacetime regions in a multiverse. We assume that the 10^{500} different possible Calabi–Yau shapes are physically realized in different universes, resulting in universes with different physical parameters – different particle spectra, different physical constants and laws.

There is no observational or experimental evidence for any of these assumptions. So, my degree of belief in h_M is virtually non-existent.

Applying Bayesian logic at this point doesn't change the picture much, if it all.

Of course, the superstring theorists will argue that this isn't a valid comparison. The multiverse theory h_M should *subsume* the single-universe theory h_U, such that h_U is seen to be an approximation of the larger theory. Fine. When the theorists are able to use the multiverse theory to calculate everything that the standard models of particle physics and big bang cosmology can calculate, I'll be ready to reconsider.

Where does this leave us? In *The Cosmic Landscape*, Susskind explained that he could find only two kinds of arguments against anthropic reasoning:

> As much as I would very much like to balance things by explaining the opposing side, I simply can't find the other side. Opposing arguments boil down to a visceral dislike of the anthropic principle (I hate it) or an ideological complaint against it (it's giving up).[21]

I'd like to offer a third argument. I don't hate the weak anthropic principle. I don't think its adoption by theorists means they've given up. I reject the weak anthropic principle because it is simply empty of scientific content. It adds absolutely nothing to the debate. And yet it is used by some contemporary theorists to provide a rather facile logic, a veneer to deflect the fact that multiverse theories themselves are not scientific. Anthropic reasoning is the last refuge of theorists desperate to find a way to justify and defend their positions.

So, what do I think is going on? How do I explain the fine-tuning of the universe? My hands are in the air. It's a fair cop. I have no explanation because *science* does not yet have an explanation. We may be here because, by happy accident or the operation of some complex natural physical mechanisms we have yet to fathom, the parameters of the universe just happen to be compatible with our existence.

And this is indeed the point. Scientists (even theoretical physicists) should not be afraid to say that they don't know. Nobody is expecting them to have all the answers to human existence. We want them to speculate, to push the frontiers of their science. But when their ambition to give answers drives them to tell fairy tales, smothered in a sugar-coating of anthropic logic, let us all be clear that we've left science far behind.

Just Six Questions

Defining the Destination at the End of a Hopeful Journey

As far as the laws of mathematics refer to reality, they are not certain; and as far as they are certain, they do not refer to reality.

Albert Einstein[1]

I've found it hard to reach this point without my head filling with all kinds of homilies about journeys and destinations. I set out with the intention to provide you with at least an entertaining journey, one that I trust you have travelled hopefully. These are serious issues, and I still hope to persuade you that there's quite a lot at stake. But I also hope I've been able to make my points in a way that you've found enjoyable, with a minimum of po-faced proselytizing.

We've now arrived at our destination, and it's incumbent on me to try to wrap things up as best I can and give you some sort of definition of precisely where we are. In thinking about how best to do this, I considered providing you with some kind of summary, but this seemed to involve simply repeating all my main points, and I'm conscious that I've done that a couple of times already.

Instead, I figured it would be more interesting to anticipate some of the questions that might now be lurking at the back of your mind, and try to answer them as best I can.

I've thought of six.

If fairy-tale physics isn't science, what is it?

This is a very good question (and thank you for asking). As I've tried to explain, when there is no longer an attempt to ground theoretical developments by forcing them to relate to the hard facts of empirical reality, the result is metaphysics. Therefore, the easy answer to this question is to declare that fairy-tale physics is metaphysics.

But this won't do. I went to great lengths in Chapter 1 to explain that reality (in its broader, non-empirical sense) is a metaphysical concept and that it is extremely difficult, if not impossible, to eliminate all the metaphysical elements from what we would have no difficulty in accepting as perfectly legitimate science. Philosophers of the logical positivist persuasion had a go at this in the 1920s and 1930s, and failed.

I'm not a professional philosopher, but it seems to me that the root of the problem lies in what we tend to demand from a description or explanation of the physical world if it is to be acknowledged as satisfying and meaningful.

Just what does it mean to comprehend or understand something? We might argue that we understand something when we have grasped the rules governing its properties and behaviour and can use these rules to manipulate the present or predict the future.

But there are rules, and then there are *rules*. We demand descriptions based on rules of the kind that convey insight and *understanding*. If we strip all the metaphysics from science – all the stuff we can't prove by reference to the facts but nevertheless believe to be contingently true (in the spirit of the Veracity Principle) – then what we are left with is a rather vacuous instrumentalism. Theories then serve a simple purpose. They are instruments that we use to relate one set of empirical facts to another, but in a way that conveys no real understanding.

I want to illustrate why I don't think this is very satisfactory by reference to a famous allegory devised by American philosopher John Searle. Searle formulated his 'Chinese room' thought experiment to attack some assumptions regarding artificial intelligence, but it will serve our purposes here to help us recognize what we mean by 'understanding'.

Searle sits on a chair in an empty room. We slip him pieces of paper beneath the door carrying questions written in Chinese which require simple, single-word answers. Searle doesn't speak Chinese, but he has a handbook titled *Empirical Rules for the Interpretation of Chinese Characters*. The rules described in the handbook tell him how to interpret the characters without providing a translation. They are rules of the type 'if 龙 then 战'. By blindly following the rules, Searle can select quite sensible answers in a way that does not contradict previous answers.

He slips his answers back to us beneath the door. From our perspective, he is having an intelligent conversation with us, in Chinese.

But Searle doesn't *understand* Chinese. He is just a computer, following an algorithm. To have an intelligent conversation with the universe, we need more than mechanical algorithms or instrumental rules that allow us to manipulate the present or predict the future. We need a *structure*: we need to understand the letters, the vowels, the words, the grammar and syntax. We need to understand the basic elements of reality – light, matter and force, space and time – and how these elements combine together to construct the poetry of the universe.

Metaphysics is an inherent and perfectly natural part of the language we use in our dialogue with nature. Eliminate it completely and the language becomes devoid of real meaning. We find we can no longer hold a sensible conversation in it.

The logical positivist programme was doomed. The philosopher A. J. Ayer, the positivist movement's English spokesman, was obliged to develop a grudging acceptance of metaphysics. He observed that: 'The metaphysician is treated no longer as a criminal but as a patient …'[2]

But even though our scientific dialogue has metaphysical elements, this is still a dialogue that is about real, empirical things. It concerns effects that we interpret in terms of the properties and behaviour of photons or quarks or electrons: their spin properties, their quantum wave-particle duality, as manifested in the observations we make and the experiments we perform. It concerns the curvature of spacetime in the vicinity of the sun. It concerns the quantum ripples in the CMB radiation. The dialogue has metaphysical elements (such as photons or quarks, or …), but of course it is also rich in empirical content.

In fairy-tale physics, we lose sight of the empirical content, almost completely. Yes, of course there are references to photons and quarks and electrons, spacetime curvature and quantum ripples, but these are broadly qualitative, not quantitative, references. And we get a lot more besides – sparticles, hidden dimensions, Kaluza–Klein particles, branes, many worlds, other universes, and so on. These new theoretical entities come also in references that are broadly qualitative, not quantitative. If there is one theme underpinning contemporary theoretical physics, it seems to be an innate inability to calculate *anything*, with the not-so-apologetic caveat: well, it still might be true.

I'm no big fan of out-and-out empiricism, but Scottish philosopher and arch-sceptic David Hume's oft-quoted passage seems particularly relevant here:

> When we run over libraries, persuaded of these principles, what havoc must we make? If we take in our hand any volume of divinity or school metaphysics, for instance, let us ask, Does it contain any abstract reasoning concerning quantity or number? No. Does it contain any experimental reasoning concerning matter of fact and existence? No. Commit it then to the flames, for it can contain nothing but sophistry and illusion.[3]

The issue, then, is not metaphysics per se. The issue is that in fairy-tale physics the metaphysics *is all there is*. Until and unless it can predict something that can be tested by reference to empirical facts, concerning quantity or number, it is nothing but sophistry and illusion.

But aren't theoretical physicists supposed to be really smart people?

Another good question. You don't get to be a professor of theoretical physics at Stanford, Harvard or Princeton, or any leading academic institution anywhere in the world, if you're an intellectual lightweight. These are *very* smart cookies. So how can I claim they've got it all wrong?

Well, obviously *they* don't think what they're doing is wrong. As far as they're concerned, this all makes perfect sense. We know the authorized version of reality can't be right. Attempts to solve its problems lead to mathematical structures that suggest all kinds of bizarre things. Okay, we have no empirical proof for these things, but we're faced with a choice. Either we give up and just admit we don't know (which is regarded by many as a rather lame response), or we push on past the demand for empirical proof and explore the structures in a lot more detail, perhaps in the hope that *something* will turn up which will allow us to connect the mathematical structures back to reality.

I actually don't have a problem with this. We should be glad that those few theorists pursuing quantum field theory in the 1950s and 1960s stuck with their programmes, against the odds that prevailed at the time. The 'bootstrap' model was the more popular theoretical structure, and the contributions of quantum field theorists were generally dismissed as irrelevant. In 1964, Peter Higgs had problems publishing the paper in which he outlined what would become known as the Higgs mechanism (and in which, in a subsequently added footnote, he predicted the existence of the Higgs boson). He later wrote:

> I was indignant. I believed that what I had shown could have important consequences in particle physics. Later, my colleague Squires, who spent the month of August 1964 at CERN, told me that the theorists there did not see the point of what I had done. In retrospect, this is not surprising: in 1964 … quantum field theory was out of fashion …[4]

Here's my problem. For how long do we continue to suspend our demand for empirical proof? Ten years? Thirty years? A hundred years? The assumption of electro-weak symmetry-breaking caused by the Higgs field allowed Steven Weinberg to predict the masses of the W and Z particles in 1967. The Higgs mechanism was therefore a theoretical device that was arguably justified when these particles were discovered at CERN in the early 1980s, almost twenty years after the theory had been written down.

But even this is indirect evidence, at best. The existence of the Higgs field can only really be betrayed by observing the Higgs boson. If the new particle discovered at CERN in July 2012 is indeed the Higgs boson, then almost fifty years have elapsed since this particle's invention.

We have obviously had to learn to be patient. But at least the Higgs mechanism was progressive – it solved some problems and made predictions for which there was some hope of providing a test sometime in the not too distant future. SUSY has made some predictions, of a sort, but these are not so far supported by data emerging from the LHC. SUSY is failing the test. Superstrings/M-theory and the various multiverse theories have made no really testable predictions at all.

At what point do we recognize that the mathematical structures we're wrestling to come to terms with might actually represent a wrong turn, like the bootstrap model?

Once we ease off on our demands, once we abandon the checks and balances afforded by a proper adherence to the scientific method, we train a whole generation (perhaps that should be generations) of theorists to believe that a soft approach to empirical proof is not only perfectly acceptable but even *necessary* to continue to publish research papers and advance their careers.

Now this is where it all gets *really* interesting. What happens when some really clever people decide that it's okay to abandon the checks and balances? After all, they say, we're all grown-ups. We're all smart cookies. According to a strict interpretation of the rules, this is maybe not acceptable. But hey, rules are made to be broken. If we just relax the rules even by a little bit, then all manner of new and exciting things become possible.

This shift in value-set can be overt or it can be quite subtle. Irrespective of its subtlety, such a shift can lead inexorably to the development of what I call a Grand Delusion (with Capital Letters, intended to Emphasize its Profound Importance).

There are plenty of precedents for Grand Delusions throughout history.* And 'cleverness', it seems to me, is almost a prerequisite. The best example of a recent Grand Delusion that I can think of is the one that led some very clever people in the global financial sector to think that they could relax the rules on financial risk, thereby discovering that all manner of new and exciting things become possible.

Lending money has always been a bit of a gambler's game. But banks and sub-prime mortgage lenders found that they could substantially grow their businesses by rewriting the rules on risk assessment and concealing the nature of this risk by salami-slicing the debt and selling it on to a multiplicity of other institutions through 'collateralized debt obligations' and other complex financial instruments.

At the heart of this new financial wizardry was Chinese mathematician David X. Li's Gaussian copula function. Li was lauded as the 'world's most influential actuary'. His function was used to model complex risk

---

* Any doubters might want to consult Charles Mackay's *Extraordinary Popular Delusions and the Madness of Crowds*, first published in 1841. Among the economic Grand Delusions Mackay lists the Dutch tulip mania of the early seventeenth century and the South Sea Bubble of 1711–20. Thanks to Professor Steve Blundell for drawing this book to my attention.

with a minimum of fuss and with apparent accuracy. It was applied by the banks and the mortgage lenders to set prices for their collateralized debt obligations.

Suddenly it became possible to lend money to institutions and individuals who would previously have been deemed high-risk borrowers. The value-set changed. Lending became more and more predatory, and the result was a property boom.

Those of us with investment savings accounts and pension funds enjoyed it while it lasted. We welcomed an unprecedented period of economic growth. But we were all conspirators in a conspiracy not of our devising. Our financial institutions applied their new theoretical structures and made hay while the sun shone. In the UK, the Labour government's Chancellor of the Exchequer Gordon Brown played whore to the financial sector, declaring the end of boom and bust. It really did seem as though the 'City' (meaning the financial centres of major cities around the world) had figured out how to make money out of money in ways that benefited many, harmed nobody and would continue for ever.

Alas. It was a Grand Delusion. There finally came a day of reckoning in October 2008. No matter how elaborate the theoretical structures or how complex the financial instruments, nothing could change the rude facts of empirical reality.* An awful lot of money had been lent to people who would never pay it back. It was as simple as that. The boom turned to bust.

We had all received a rather stark lesson in the cost of allowing mathematical models to run too far ahead of the empirical facts: 'Li's Gaussian copula formula will go down in history as instrumental in causing the unfathomable losses that brought the world financial system to its knees.'[5]

Now, I don't wish to underestimate the intellectual capabilities of theoretical physicists, who, I'm sure, are a lot smarter than actuaries, bankers and mortgage lenders. But it does seem to me that if a relatively small number of very smart people in the financial sector can delude themselves in a way that almost brought down the entire world economy, and which four years later still threatens to cause some

---

* In this case, social rather than physical reality – see my book *A Beginner's Guide to Reality*.

European countries to default on their sovereign debts, then it's surely possible that a few theorists can delude themselves about what qualifies as science?

Okay, but in the grand scheme of things is there any real harm done?

Of course, arguing about whether M-theory, the multiverse and other products of fairy-tale physics are metaphysical rather than scientific structures offers something of an entertaining distraction, but you might think this is hardly earth-shattering when measured (for example) against the misuse of a mathematical function to price collateralized debt obligations. After all, what does it matter if a few theorists decide that it's okay to indulge in a little self-delusion? So what if they continue to publish their research papers and their popular science articles and books? So what if they continue to appear in science documentaries, peddling their metaphysical world views as science? What real harm is done?

I believe that damage *is* being done to the integrity of the scientific enterprise. The damage isn't always clearly visible and is certainly not always obvious. Fairy-tale physics is like a slowly creeping yet inexorable dry rot. If we don't look for it, we won't notice that the foundations are being undermined until the whole structure comes down on our heads.

Here are the signs.

The fairy-tale theorists have for some time been presenting arguments suggesting that the very definition of science needs to be adapted to accommodate their particular brand of metaphysics. The logic is really rather simple. Developments in theoretical physics have run far ahead of our ability to provide empirical tests. If we hang our definition of science on the Testability Principle, then we have a problem – this stuff clearly isn't science.

But if it isn't science, we forgo the opportunity to explore the vast richness afforded by the mathematics. This richness is compelling, as it feeds an innate human desire to concoct stories about our universe and our place in it. Something has to give.

So, in *The Hidden Reality*, Greene writes:

Sometimes science does something else. Sometimes it challenges us to reexamine our views on science itself. The usual centuries-old scientific framework envisions that when describing a physical system, a physicist needs to specify three things ... First are the mathematical equations describing the relevant physical laws ... Second are the numerical values of all constants of nature that appear in the mathematical equations ... Third, the physicist must specify the systems' initial conditions ... The equations then determine what things will be like at any subsequent time ... Yet, when it comes to describing the totality of reality, the three steps invite us to ask deeper questions: Can we explain the initial conditions – how things were at some purportedly earliest moment? Can we explain the values of the constants – the particle masses, force strengths, and so on – on which these laws depend? Can we explain why a particular set of mathematical equations describes one or other aspect of the physical universe?[6]

This all seems very reasonable. But at what point does asking 'deeper questions' about the 'totality of reality' take us across the threshold from physics to metaphysics? At what point do we run up against the limits of science's capability to answer these deeper questions?

To be fair, the history of science is a history of continually and consistently pushing the boundaries into domains in which it might have been thought that science could or should have no answers to give. To the surprise of many, there were indeed answers to be found. And these were answers that were found to be rooted in empirical reality.

But in fairy-tale physics there are no answers, there are only untested and untestable speculations. Greene continues:

Collectively, we see that the multiverse proposals ... render prosaic three primary aspects of the standard scientific framework that in a single-universe setting are deeply mysterious. In various multiverses, the initial conditions, the constants of nature and even the mathematical laws are no longer in need of explanation.[7]

Don't get me wrong. I'm all in favour of pushing the boundaries; of probing the limits of scientific enquiry. I'm fundamentally interested in answers to deeper questions about the totality of reality. But I personally prefer rigorous answers that *mean* something. I prefer answers that I can regard either as scientifically true or capable in principle of yielding answers that might one day come to be regarded as true.

Greene wants to give me these answers (basically, telling me that in the anthropic multiverse no answers are necessary, or all answers are possible) by insisting that we extend the definition of science to include speculative theorizing of the fairy-tale kind. We must give up our checks and balances. We must abandon the Testability Principle and adapt our understanding of what it means for something to be scientifically true.

Susskind argues along similar lines. In *The Cosmic Landscape*, he writes:

> Frankly, I would have preferred to avoid the kind of philosophical discourse that the anthropic principle excites. But the pontification, by the 'Popperazzi', about what is and what is not science has become so furious in news reports and internet blogs that I feel I have to address it. My opinion about the value of rigid philosophical rules in science is the same as Feynman's ['Philosophers say a great deal about what is absolutely necessary for science, and it is always, so far as one can see, rather naïve, and probably wrong.'][8]

The term 'Popperazzi' is a reference to the Austrian philosopher Karl Popper, whose principle of falsifiability continues to be popular among scientists as a working definition for what constitutes a scientific theory. According to this view, a scientific theory is one that is in principle capable of being falsified by an observation or an experimental test. Unfortunately, falsifiability has long since lost credibility among philosophers of science.

In fact, 'Popperazzi' is also a not-so-veiled reference to theorist Lee Smolin, who has sought to use Popper's principle of falsifiability to argue that the anthropic principle is not science. Susskind and Smolin have publicly debated these issues, for the most part arguing at cross-purposes.[9]

It is unfortunate that Smolin used falsifiability as a criterion with which to question the scientific authenticity of the anthropic principle, since falsifiability is open to attack and not easily defended. But at the heart of Susskind's argument there appears to lurk a reluctance to commit to 'rigid rules', no matter what their origin.

By the way, I think that Susskind's use of Feynman's words in support of his argument is rather disingenuous. Feynman was no great fan of philosophy, but he understood well enough what science is. He was also extremely sceptical of the string theory approach. Another oft-quoted Feynman observation is that: 'String theorists don't make predictions, they make excuses.'[10]

I think we have to be quite concerned about where these attempts to change our outlook on science might lead. Softening our position and letting in all manner of speculative metaphysics may give some theorists a stronger sense of justification for what they're doing, but what other consequences might follow?

If we're going to abandon the 'rigid rules' and change the definition of what it means for something to be scientifically true, then what definition should we use? That it is rigorously mathematically consistent and 'coherently' true? But why stop there? If we're no longer demanding that scientific theories should establish a correspondence truth with elements of empirical reality, then, surely, anything goes? Why wouldn't we then regard astrology as true? Homeopathic medicine? Intelligent design?

We can see where this leads. If scientists can set themselves up as the high priests of a new metaphysics, and continue to preach their gospel unchallenged through popular books and television, then the credibility of all scientists starts inexorably to be eroded. Why should we take any of them seriously?

The really worrying thing is that the scientific community seems caught in two minds about all this. While there are many physicists prepared to take the tellers of fairy tales to task, this is extremely sensitive ground. It is hard to criticize fairy-tale physics without being perceived to be criticizing science as a whole. And at a time when science is on the defensive against a resurgent and increasingly voluble anti-science rhetoric, is it really helpful to be throwing rocks at a few theorists? With the boat already rocking, physicists think twice before jumping up and down in it.

What do the philosophers have to say about all this?

Hmm. Another interesting question. We might have imagined that these rather striking developments in theoretical physics would have attracted a lot of interest from the philosophy community. After all, the fairy-tale physicists appear to be challenging the very definition of science, and this is something on which philosophers of science might have been expected to have a ready opinion.

In fact, almost a century of intellectual endeavour and argumentation appears to have led the philosophers further and further away from a consensus on science and the scientific method. As Greek philosopher Stathis Psillos, an authority on scientific realism, commented in a personal note to me:

> I share your concerns [regarding] the [current] state of [the] philosophy of science. There is nothing like a consensus on scientific method, explanation, causation, laws and all other key concepts and issues in the philosophy of science. Actually, it seems that if there is anything like a growing tendency it is for pluralism. When it comes to the metaphysics of science, it seems the tendency is to go back to Aristotelianism – and here is where things go bizarre really, since the connection with the current scientific image of the world is thinner than ever.[11]

In broad terms, the pluralism that Psillos mentions refers to the very different perspectives held by contemporary philosophers of science. Some, like Psillos himself, argue for a form of scientific realism. Others favour a form of post-positivist empiricism, retaining the negative attitude towards metaphysics and denying that scientific theories progress towards a literally true representation of an independently existing reality. The empiricist believes that the purpose of science is rather to provide us with theories that are adequate for the task of relating one set of facts to another. We should accept as true only what these theories have to say about those things that we can see directly for ourselves. But we should not believe that the unobservable entities that the theories describe (such as photons, quarks or electrons) are in themselves real and that the theories are literally true.

Yet others argue that scientific theories are social constructions, that their interpretation and acceptance as 'the truth' are no more than conventions, achieved through consensus within the community of people engaged in scientific activity in any particular generation. For sure, these constructions might collectively offer a more *reliable* interpretation of nature than those afforded by superstition or religious mythology, but they are constructions nonetheless.

Steven Weinberg was expressing exasperation with this state of affairs as early as 1993:

> This is not to deny all value to philosophy, much of which has nothing to do with science. I do not even mean to deny all value to the philosophy of science, which at its best seems to me a pleasing gloss on this history and discoveries of science. But we should not expect it to provide today's scientists with any useful guidance about how to go about their work or about what they are likely to find.[12]

Susskind goes even further. He believes that the guardians of science and scientific methodology are scientists, not philosophers:

> Good scientific methodology is not an abstract set of rules dictated by philosophers. It is conditioned by, and determined by, the science itself and the scientists who create the science. What may have constituted scientific proof for a particle physicist of the 1960s – namely the detection of an isolated particle – is inappropriate for a modern quark physicist who can never hope to remove and isolate a quark. Let's not put the cart before the horse. Science is the horse that pulls the cart of philosophy.[13]

Despite this wave of general negativity, if not outright hostility, towards philosophers, there are a few (rare) instances in which philosophers have deigned to pass judgement. For example, in September 2007, philosophers Nancy Cartwright and Roman Frigg provided a short commentary on string theory which was published in the British science monthly *Physics World*.

In my view, Cartwright and Frigg were actually rather generous in their assessment of string theory's 'successes', but they were quite clear about the theory's lack of 'progression':

> The question of how progressive string theory is then becomes one of truth, and this brings us back to predictions. The more numerous, varied, precise and novel a theory's successful predictions are, the more confidence we can have that the theory is true, or at least approximately true ... That a theory describes the world correctly wherever we have checked provides good reason to expect that it will describe the world correctly where we have not checked. String theory's failure to make testable predictions therefore leaves us with little reason to believe that it gives us a true picture.[14]

I think it's high time we heard a bit more from the philosophers. I'd be interested in their interpretation of the status of M-theory, the multiverse and the anthropic principle. A hostile reception can be pretty much guaranteed, but I believe it is vitally important that the guardianship of science and the scientific method should not be left solely in the hands of scientists, particularly those scientists with intellectual agendas of their own.

Are we witnessing the end of physics?

In the years building up to the end of the last millennium, a considerable stir was caused by a number of popular books declaring the 'end' of things. The trend may have begun with Francis Fukuyama's *The End of History and the Last Man*, which was published in 1993. In 1996, John Horgan, then a staff writer at *Scientific American*, weighed in with *The End of Science*.

Horgan took a generally pessimistic view, not just of physics and cosmology, but of philosophy, evolutionary biology, social science and neuroscience. *The End of Science* was generally rather destructive and unhelpful, which was a shame, because Horgan did have some really valuable points to make. He reserved particular ire for the state of contemporary physics:

This is the fate of physics. The vast majority of physicists, those employed in industry and even academia, will continue to apply the knowledge they already have in hand – inventing more versatile lasers and superconductors and computing devices – without worrying about any underlying philosophical issues. A few diehards dedicated to truth rather than practicality will practice physics in a nonempirical, ironic mode, plumbing the magical realm of superstrings and other esoterica and fretting about the meaning of quantum mechanics. The conferences of these ironic physicists, whose disputes cannot be experimentally resolved, will become more and more like those of that bastion of literary criticism, the Modern Language Association.[15]

In truth, we've been here before. Although entirely apocryphal,[16] we shouldn't let this get in the way of a neat story, and the story goes that in 1900,* the great British physicist Lord Kelvin famously declared to the British Association for the Advancement of Science that: 'There is nothing new to be discovered in physics now. All that remains is more and more precise measurement.'[17] What followed was a century of scientific discovery on an unprecedented scale.

This tends to be the fate of any declaration that we've reached the end. Horgan's claim that we'd reached the end of physics and cosmology was swiftly followed in 1998 by the announcement – to the considerable astonishment of theorists – that the expansion of the universe is actually accelerating. Leggett demonstrated that it was still possible to explore the meaning of quantum mechanics through experiment, and experiments were duly performed and reported in 2007.**

And although Horgan might be tempted to dismiss the discovery of the electro-weak Higgs boson in 2012 (if this indeed is what it is) as something that was predicted nearly fifty years ago and so takes particle physics no further forward, the simple truth is that this is a particle we know relatively little about. If this is the Higgs boson, then there is much we still need to learn about its properties and behaviour. Now

---

* There's obviously something about the turning of centuries that brings out this kind of stuff.

** Although, in fairness, the experiments only served to deepen the mystery even further.

that we can make it, we have an opportunity to study it in great detail. There may yet be surprises in store.

We're far from the end of physics and cosmology. There is still a long way to go and we have a long list of unanswered questions. But there is an inescapable consequence of the stage of maturity that physics has reached. Discoveries of significant new empirical facts are now very few and very far between. In the meantime, our Western scientific–technical culture has developed a seemingly insatiable appetite for instant gratification. We want answers *now*. Theorists need to attend conferences and publish papers *this year*. They (and we) don't want to be kept waiting another two decades (three? five?) for the next discovery, the next big empirical fact.

In this sense Horgan got it exactly right, although he might have emphasized that the most important psychological factor driving the development of what he called 'ironic physics' (and what I've called fairy-tale physics) is childish *impatience*.

I think he also underestimated just how esoteric the esoterica would become.

So, what do you want *me* to do about it?

This is easily answered, and you'll find I'm not very demanding. There are already some signs that the grip of the fairy-tale physicists may be weakening. Failed theories don't get junked overnight. They tend to fade away none-too-gracefully, as theorists gradually realize that their time and energy may be more fruitfully spent on other things. The annual string theory conference, last held in June 2011, was less well attended than previous events, although this might have been simply because the cost of attending this conference (in Uppsala, Sweden) was quite high and videos of the talks were freely available online.[18]

The absence of hard evidence from the LHC for sparticles and other phenomena tentatively 'predicted' by SUSY and superstring theories is greatly discouraging, despite the tendency of SUSY and string phenomenologists to put a brave face on things. My own feeling is that the interconnectedness of all the unjustified and untestable assumptions that have been deployed in the creation of fairy-tale physics will slowly but inevitably bring the whole structure down. Future theorists may look back at this period in the development of physics and wonder why so few thought to challenge the orthodoxy of the time. Didn't we appreciate that something funny was going on?

My hope is that in its exploration of what looks likely to be the electro-weak Higgs boson, the various detector collaborations at the LHC turn up some really puzzling new facts. No doubt the theorists will be quick to explain how any new facts are consistent with SUSY, M-theory or the multiverse, but I'm reasonably confident that common sense will ultimately prevail.

In the meantime, we have to square up to the challenge posed by fairy-tale physics. And this is all I ask of you. Next time you pick up the latest best-selling popular science book, or tune into the latest science documentary on the radio or television, keep an open mind and try to maintain a healthy scepticism. By all means allow yourself to be entertained, but remember Hume's quote above. What is the nature of the evidence in support of this theory? Does the theory make predictions of quantity or number, of matter of fact and existence? Do the theory's predictions have the capability – even in principle – of being subject to observational or experimental test?

Come to your own conclusions.

Endnotes

Quotes and other references to texts listed in the Bibliography are indicated by author surname, title (where necessary) and page number. Most of the Einstein quotes used in the chapter headings can be found in Alice Calaprice, *The Ultimate Quotable Einstein* (Princeton University Press, 2011). There are several references to articles posted on the online preprint archive arXiv, managed by Cornell University. These can be accessed by loading the arXiv home page – http://arxiv.org/ – and typing the article identifier in the search window.

Chapter 1: The Supreme Task

1   Albert Einstein, 'Motives for Research', a speech delivered at Max Planck's sixtieth birthday celebration, April 1918.
2   Larry and Andy Wachowski, *The Matrix: The Shooting Script*, Newmarket Press, New York, 2001, p.38.
3   Bernard d'Espagnat, *Reality and the Physicist: Knowledge, Duration and the Quantum World*, Cambridge University Press, 1989, p.115.
4   From the 1978 essay 'How to Build a Universe that Doesn't Fall Apart Two Days Later', included in the anthology *I Hope I Shall Arrive Soon*, edited by Mark Hurst and Paul Williams, Grafton Books, London, 1988. This quote appears on p.10.
5   Heisenberg, p.46.
6   Quoted in Maurice Solovine, *Albert Einstein: Lettres à Maurice Solovine*, Gauthier-Villars, Paris, 1956. This quote is reproduced in Fine, p.110.
7   Hacking, p.23.
8   Ian Sample, 'What is this thing we call science? Here's one definition …', *Guardian*, 4 March 2009. www.guardian.co.uk/science/blog/2009/mar/03/science-definition-council-francis-bacon. More details can be found on the Science Council's website: www.sciencecouncil.org/content/what-science.

ENDNOTES

9   Jon Butterworth, 'Told You So … Higgs Fails to Materialise', *Life and Physics*, hosted by the *Guardian*, blog entry, 11 May 2011. www.guardian.co.uk/science/life-and-physics.

10   Pierre Duhem, *The Aim and Structure of Physical Theory*, English translation of the second French edition of 1914 by Philip P. Wiener, Princeton University Press, 1954, p.145.

11   Johannes Kepler, *Astronomia Nova*, Part II, Section 19, quoted in Koestler, pp.326–7.

12   Ibid., Part IV, Section 58, quoted in Koestler, p.338.

13   Letter to Richard Bentley, quoted in Koestler, p.344. The letter appears to be undated but is a reply to one from Bentley dated 18 February 1693. Transcripts of both letters can be viewed online at The Newton Project, www.newtonproject.sussex.ac.uk.

14   Russell, p.34.

15   Ibid, p.35.

Chapter 2: White Ambassadors of Morning

1   Letter to Heinrich Zangger, 20 May 1912.

2   Pink Floyd, *Meddle*, originally released on Harvest Records (an imprint of EMI), October 1971.

3   Albert Einstein, *Annalen der Physik*, 17 (1905), p.133. English translation in Stachel, p.178.

4   Following Planck, Einstein suggested that the energy (E) carried by an individual light quantum is proportional to its frequency (given by the Greek symbol nu, ν), according to $E = h\nu$, where h is Planck's constant, the fundamental *quantum of action* in quantum mechanics. In many ways, this result rivals his most famous equation $E = mc^2$, which we will encounter in Chapter 4.

5   Quoted in Pais, *Subtle is the Lord*, p.382.

6   Quoted by Aage Petersen in French and Kennedy, p.305.

7   This is given as the quantum number multiplied by Planck's constant h divided by 4π.

8   Letter to Max Born, 4 December 1926. Quoted in Pais, *Subtle is the Lord*, p.443.

9   The two are related according to the equation $\lambda = v/\nu$, where the Greek symbol lambda, λ, represents the wavelength, v is the phase velocity and ν is the frequency of the wave.

10   The de Broglie relationship is written $\lambda = h/p$, where λ is the wavelength, h is Planck's constant and p is the momentum.

11   Heisenberg established that the uncertainty in position multiplied by the uncertainty in momentum must be greater than, or at least equal to, Planck's constant h.

12   Albert Einstein, Boris Podolsky, Nathan Rosen, *Physical Review*, 47, 1935, pp.777–80. This paper is reproduced in Wheeler and Zurek, p.141.

13   In Stefan Rozenthal (ed.), *Niels Bohr: His Life and Work as Seen by his Friends and Colleagues,* North-Holland, Amsterdam, 1967, pp.114–36. An extract of this essay is reproduced in Wheeler and Zurek, pp.137 and 142–3. This quote appears on p.142.

14   Paul Dirac, interview with Niels Bohr, 17 November 1962, Archive for the History of Quantum Physics. Quoted in Beller, p.145.

15   A. J. Leggett, *Foundations of Physics*, 33 (2003), pp.1474–5.

16   John Bell, *Epistemological Letters*, November 1975, pp.2-6. This paper is reproduced in Bell, pp.63–6. The quote appears on p.65.

17   The expression 2.697±0.015 indicates the spread of experimental results around the mean value. The experiments produced results in the range 2.682 to 2.712, representing 68 per cent confidence limits or one standard deviation.

18   See Xiao-song Ma, et al., arXiv: quant-ph/1205.3909v1, 17 May 2012.

19   Why do all these experiments never quite achieve the precise quantum theory predictions? Because entangled quantum particles are easily 'degraded'. Entangled particles which lose their correlation because of stray interactions or instrumental deficiencies will look, to all intents and purposes, as though they are locally real and will not contribute to a violation of the inequality being measured. The wavefunctions of such entangled particles have been prematurely collapsed.

20   Niels Bohr, *Physical Review*, 48 (1935), pp.696–702. This paper is reproduced in Wheeler and Zurek, pp.145–51. The quote appears on p.145 (emphasis in the original).

Chapter 3: The Construction of Mass

1   Albert Einstein, *Autobiographical Notes*, 33 (1946).

2   Remember, the number of different spin orientations is given by twice the spin quantum number plus 1. For s = ½, this gives 2 × ½ + 1, or two spin orientations.

3   In fact, Einstein's famous equation $E = mc^2$ is an approximation of a more complex equation $E^2 = p^2c^2 + m^2c^4$, where p is the momentum and m is the rest mass. In any relativistic treatment, energy must therefore enter as E^2 and there will always be two sets of solutions, one with positive energy and one with 'negative' energy.

4   Freeman Dyson, *From Eros to Gaia*, Pantheon, New York, 1992, p.306. Quoted in Farmelo, *The Strangest Man*, p.336.

5   Feynman, p.188. This quote appears in the caption to Figure 76.

6   Quoted in Kragh, *Quantum Generations*, p.204.

7   Willis Lamb, *Nobel Lectures, Physics 1942–1962*, Elsevier, Amsterdam, 1964, p.286.

8   Quoted by Kragh, *Quantum Generations*, as 'physics folklore', p.321.

9   Nambu, p.180.

10   Steven Weinberg, *Nobel Lectures, Physics 1971–1980*, edited by Stig Lundqvist, World Scientific, Singapore, 1992, p.548.

11   Interview with Robert Crease and Charles Mann, 7 May 1985. Quoted in
      Crease and Mann, p.245.

12   Our friends at the Particle Data Group list the mass of the W particles as
      80.399±0.023 GeV, or 85.713 times the mass of a proton, and the mass of the
      Z^0 as 91.1876±0.0021 GeV, or 97.215 times the mass of a proton.

13   Rolf Heuer, 'Latest update in the search for the Higgs boson', CERN
      Seminar, 4 July 2012.

14   The particle 'consistent' with the Higgs boson was found to have a mass
      around 125–6 GeV.

15   CERN press release, 4 July 2012.

16   Interview with Robert Crease and Charles Mann, 3 March 1983. Quoted in
      Crease and Mann, p.281.

Chapter 4: Beautiful Beyond Comparison

1   Letter to Heinrich Zangger, 26 November 1915. The colleague in question
     was German mathematician David Hilbert, who was in pursuit of the general
     theory of relativity independently of Einstein. See Isaacson.

2   Isaac Newton, *Mathematical Principles of Natural Philosophy*, first American
     edition, translated by Andrew Motte, published by Daniel Adee, New York,
     1845, p.81.

3   Of course, this doesn't mean that the flashlight is really 'stationary'. The
     flashlight is spinning along with the earth's rotation on its axis and moving
     around the sun at about 19 miles per second. The solar system is moving
     towards the constellation Hercules at about 12 miles per second and drifting
     upwards, above the plane of the Milky Way, at about 4 miles per second. The
     solar system is also rotating around the centre of the galaxy at about 124 miles
     per second. Add this together, and we get a speed of about 159 miles per
     second. This is the speed with which the earth is moving within our galaxy.
     Now if we consider the speed with which the galaxy moves through the
     universe …

4   Albert Einstein, *Annalen der Physik*, 17 (1905), pp.891–921. An English
     translation of this paper is reproduced in Stachel, pp.123–60. This quote
     appears on p.124.

5   The relationship can be worked out with the aid of a little high-school
     geometry. The time interval on the moving train is equal to the time interval
     on the stationary train divided by the factor $\sqrt{1 - \frac{v^2}{c^2}}$, where v is the speed of the
     train and c is the speed of light. If v = 100,000 miles per second and c =
     186,282 miles per second, then this factor has the value 0.844. This means
     that the time on the moving train is dilated by about 19 per cent. So, 24.1
     billionths of a second becomes 28.6 billionths of a second.

6   Actually, this experiment is complicated by the fact that, when transported at
     an average 10 kilometres above sea level, an atomic clock actually runs *faster*
     because gravity is weaker at this height above the ground. In the experiment
     described here, this effect was expected to cause the travelling clock to gain
     53 billionths of a second, offset by 16 billionths of a second due to time

ENDNOTES

dilation. The net gain was predicted to be about 40 billionths of a second. The measured gain was found to be 39±2 billionths of a second.

7   Again, the GPS system requires two kinds of corrections – the speed of the satellites causes a time dilation of seven thousandths of a second and the effects of weaker gravity at the orbiting distance of 20,000 kilometres from the ground causes the atomic clocks to run faster by about 45 thousandths of a second. The net correction is therefore 38 thousandths of a second.

8   The length of the train contracts by a factor $\sqrt{1 - \frac{v^2}{c^2}}$ where, once again, v is the speed of the train and c is the speed of light. If the train is moving at 100,000 miles per second and the speed of light is taken to be 186,282 miles per second, then this factor is 0.844.

9   Albert Einstein, *Annalen der Physik*, 18 (1905), pp.639–41. An English translation of this paper is reproduced in Stachel, pp.161–4. This quote appears on p.164.

10 Poincaré's paper is cited by Lev Okun, *Physics Today*, June 1989, p.13.

11 See note 3 in Chapter 3, above. The full equation is $E^2 = p^2c^2 + m^2c^4$, where p is the momentum of the object, m is its 'rest mass' and c is the speed of light. For massless photons with m = 0, the equation reduces to $E^2 = p^2c^2$, or E = |p|c, where |p| represents the modulus (absolute value) of the momentum. Photons have no mass but they do carry momentum. You might be tempted to apply the classical non-relativistic expression for momentum – mass times velocity – and so conclude that for photons |p| = mc, so $E = mc^2$ after all. But this is going around in circles. For photons m = 0, so we would be forced to conclude that photons have no momentum. What this means is that the non-relativistic expression for momentum does not apply to photons.

12 The equation is $M = m/\sqrt{1 - \frac{v^2}{c^2}}$, where M represents the 'relativistic mass', m the 'rest mass', v is the speed of the object and c is the speed of light. This seems to suggest that a passenger with a rest mass of 60 kg travelling at 100,000 miles per second would acquire a relativistic mass of about 71 kg. However, there are caveats – see text.

13 Letter to Lincoln Barnett, 19 June 1948. A facsimile of part of this letter is reproduced, together with an English translation, in Lev Okun, *Physics Today*, June 1989, p.12.

14 Hermann Minkowski, 'Space and Time', in Hendrik A. Lorentz, Albert Einstein, Hermann Minkowski and Hermann Weyl, *The Principle of Relativity: A Collection of Original Memoirs on the Special and General Theory of Relativity*, Dover, New York, 1952, p.75.

15 Greene, *The Fabric of the Cosmos*, p.51.

16 Newton, *Mathematical Principles*, op cit., p.506.

17 Albert Einstein, 'How I Created the Theory of Relativity', lecture delivered at Kyoto University, 14 December 1922, translated by Yoshimasa A. Ono, *Physics Today*, August 1982, p.47.

18 Wheeler, with Ford, p.235.

19 In A. J. Knox, Martin J. Klein and Robert Schulmann (editors), *The Collected Papers of Albert Einstein, Volume 6, The Berlin Years: Writings 1914–1917*, Princeton University Press, 1996, p.153.

ENDNOTES

Chapter 5: The (Mostly) Missing Universe

1   Albert Einstein, *Proceedings of the Prussian Academy of Sciences*, 142 (1917). Quoted in Isaacson, p.255.

2   This is commonly known as Olbers' paradox, named for nineteenth-century German amateur astronomer Heinrich Wilhelm Olbers. If the universe were really static, eternal, homogeneous and infinite, then it is relatively straightforward to show that although distant stars are dimmer, the fact that there are many more of them should mean that their total brightness does not diminish with distance. Consequently, the night sky would be expected to be ablaze with starlight. Recent investigations suggest that history has been rather kind to Olbers, and that others, such as sixteenth-century English astronomer Thomas Digges, deserve rather more credit for articulating this paradox. See Edward Harrison, *Darkness at Night: A Riddle of the Universe*, Harvard University Press, 1987.

3   Isaac Newton, *Mathematical Principles of Natural Philosophy*, first American edition, translated by Andrew Motte, published by Daniel Adee, New York, 1845, p.504.

4   George Gamow, *My World Line: An Informal Autobiography*, Viking Press, New York, 1970, p.149. Quoted in Isaacson, pp.355–6.

5   Albert Einstein, *Zeitschrift für Physik*, 16 (1923), p. 228

6   Hubble's law can be expressed as $v = H^0 D$, where v is the velocity of the galaxy, H^0 is Hubble's constant for a particular moment in time and D is the so-called 'proper distance' of the galaxy measured from the earth, such that the velocity is then given simply as the rate of change of this distance. Although it is often referred to as a 'constant', in truth the Hubble parameter H^0 varies with time depending on assumptions regarding the rate of expansion of the universe. Despite this, the age of the universe can be roughly estimated as $1/H^0$. A value of H^0 of 70 kilometres per second per megaparsec (2.3×10^{-18} per second) gives an age for the universe of 43×10^{16} seconds, or 13.6 billion years.

7   On submitting a paper describing their calculations to the journal *Physical Review*, Gamow added the name of fellow émigré physicist Hans Bethe to the list of authors. Bethe had not been involved in the work but Gamow, the author of the successful Mr Thompkins series of popular science books, had a reputation as a prankster. The possibilities afforded by a paper authored by Alpher, Bethe and Gamow had captured his imagination. Inevitably, it became known as the alpha-beta-gamma paper. The paper was published in 1948, on April Fool's Day. Gamow had originally marked it to indicate that Bethe was author *in absentia*, but the journal editor had removed this note. Bethe (who, as it turned out, was asked to review the manuscript) didn't mind. 'I felt at the time that it was rather a nice joke, and that the paper had a chance to be correct, so that I did not mind my name being added to it.' (Quoted by Ralph Alpher and Robert Herman, *Physics Today*, August 1988, p.28.) However, Alpher was not overly impressed. The subject of the paper was his doctoral dissertation. Both Gamow and Bethe were established

physicists with international reputations. Anyone reading the paper would likely conclude that these more esteemed physicists had done all the work.

8   Fred Hoyle, *The Nature of the Universe*, BBC Third Programme, 28 March 1949. A transcript of this broadcast was subsequently published in *The Listener* in April 1949. This quote is taken from Hoyle's original manuscript, selected pages of which are available to view online at http://www.joh.cam.ac.uk/library/special_collections/hoyle/exhibition/radio/.

9   Ralph Alpher and Robert Herman, *Physics Today*, August 1988, p.26.

10 Quoted by Overbye, p.130.

11 Quoted by David Wilkinson, 'Measuring the Cosmic Microwave Background Radiation', in P. James, E. Peebles, Lyman A. Page Jr and R. Bruce Partridge (eds.), *Finding the Big Bang*, Cambridge University Press, 2009, p.204.

12 Quoted by Overbye, p.237.

13 Guth, p.176.

14 J. P. Ostriker and P. J. E. Peebles, *Astrophysical Journal*, 186 (1973), p.467.

15 Quoted by Panek, p.240.

Chapter 6: What's Wrong with this Picture?

1   Albert Einstein, 'Induction and Deduction in Physics', *Berliner Tageblatt*, 25 December 1919.

2   Why 'modulus square'? The modulus of a number is its absolute value (its value irrespective of its sign – positive or negative). We use the modulus-square instead of the square of the amplitude because the amplitude itself may be a complex number (containing i, the square root of -1) but, almost by definition, the probability derived from it must be a positive real number – it refers to something measurable in the real world. The modulus-square of a complex number is the number multiplied by its *complex conjugate*. For example, if the amplitude is $0.5i$, the square of this is $0.5i \times 0.5i = -0.25$ (since $i \times i = -1$), which suggests a negative probability of -25%. However, the modulus-square is $0.5i \times 0.5(-i) = +0.25$ (since $i \times i = 1$), suggesting a positive probability of 25%.

3   Letter to Albert Einstein, 19 August 1935. Quoted in Fine, pp.82–3.

4   John Bell, *Physics World*, 3 (1990), p.34.

5   In Davies and Brown, p.52.

6   Albert Einstein, 'On the method of Theoretical Physics', Herbert Spencer Lecture, Oxford, 10 June 1933.

7   Lederman, p.363.

8   The Planck scale is a mass-energy scale with a magnitude around 10^{19} GeV, where the quantum effects of gravity are presumed to be strong. It is characterized by measures of mass, length and time that are calculated from three fundamental constants of nature: the gravitational constant, G, Planck's constant h divided by 2π, written $\hbar$ (pronounced 'h-bar') and the speed of light, c. The Planck mass is given by $\sqrt{hc/G}$ and has a value around 1.2×10^{19} GeV, or 1.2×10^{28} electron volts. The Planck length is given by $\sqrt{Gh/c^3}$ and

has a value around 1.6×10^{-35} metres. The Planck time is given by $\sqrt{G\hbar/c^5}$ (the Planck length divided by c), and has a value around 5×10^{-44} seconds.

9   John Irving, *A Prayer for Owen Meany*, Black Swan, 1990, pp.468–9.
10 Albert Einstein, *Preussische Akademie der Wissenschaften (Berlin) Sitzungsberichte*, 1916, p.688. Quoted in Gennady E. Gorelik and Viktor Ya. Frenkel, *Matvei Petrovich Bronstein and Soviet Theoretical Physics in the Thirties*, Birkhauser, Verlag, Basel, 1994, p.86.

Chapter 7: Thy Fearful Symmetry

1   Letter to Hans Reichenbach, 30 June 1920.
2   For the incurably curious, U(1) is the unitary group of transformations of one complex variable.
3   SU(2) and SU(3) are special unitary groups of transformations of two and three complex variables, respectively.
4   Interview with Robert Crease and Charles Mann, 29 January 1985. Quoted in Crease and Mann, p.400.
5   Stephen P. Martin, 'A Supersymmetry Primer', version 6, arXiv: hep-ph/9709356, September 2011, p.5.
6   Kane, pp.53, 63.
7   Martin, 'A Supersymmetry Primer', op cit., p.5.
8   Woit, pp.173–4.
9   Kane, p.67.
10 Randall, *Warped Passages*, p.269.

Chapter 8: In the Cemetery of Disappointed Hopes

1   Letter to Heinrich Zangger, 27 February 1938.
2   Letter to Theodor Kaluza, 21 April 1919. Quoted in Pais, *Subtle is the Lord*, p.330.
3   Letter to Paul Ehrenfest, 3 September 1926. Quoted in ibid., p.333.
4   Leonard Susskind, *The Landscape: A Talk with Leonard Susskind*, www.edge. org., April 2003.
5   Interview with Sara Lippincott, 21 and 26 July 2000, Oral History Project, California Institute of Technology Archives, 2002, p.17.
6   Interview with Sara Lippincott, 21 and 26 July, 2000, ibid., p.26.
7   Woit, pp.173–4.
8   Interview with Shing-Tung Yau, 7 February 2007. Quoted in Yau and Nadis, pp.131–2.
9   Michael Duff, 'A Layman's Guide to M-theory', arXiv: hep-th/9805177v3, 2 July 1998.
10 Kragh, *Higher Speculations*, p.303.
11 Quoted by Randall, *Warped Passages,* p.304.
12 Veltman, p.308.
13 Sheldon Glashow and Ben Bova, *Interactions: A Journey Through the Mind of a Particle Physicist*, Warner Books, New York, 1988, p.25.

ENDNOTES

14 Gordon Kane, 'String Theory and the Real World', *Physics Today*, November 2010, p.40.
15 Yau and Nadis, pp.224–5.
16 In P. C. W. Davies and Julian Brown, eds., *Superstrings: A Theory of Everything*, Cambridge University Press, 1988, p.194.
17 Randall, *Warped Passages*, jacket copy.
18 Greene, *The Fabric of the Cosmos*, jacket copy.
19 Hawking and Mlodinow, p.181.
20 Quoted by John Matson, *Scientific American*, 9 March 2011.

Chapter 9: Gardeners of the Cosmic Landscape

1 Albert Einstein, 'On the Generalised Theory of Gravitation', *Scientific American*, April 1950, p.182.
2 'Have you noticed that Bohm believes (as de Broglie did, by the way, 25 years ago) that he is able to interpret the quantum theory in deterministic terms? That way seems too cheap to me.' Letter to Max Born, 1952. Quoted in John S. Bell, *Proceedings of the Symposium on Frontier Problems in High Energy Physics*, Pisa, June 1976, pp.33–45. This paper is reproduced in Bell, pp.81–92. The quote appears on p.91.
3 H. D. Zeh, *Foundations of Physics*, 1 (1970), pp.69–76.
4 These estimates are taken from Roland Omnès, *The Interpretation of Quantum Mechanics*, Princeton University Press, 1994. The original calculations were reported in E. Joos and H. D. Zeh, *Zeitschrift für Physik*, B59 (1985), pp.223–43.
5 John S. Bell, 'Against Measurement', *Physics World*, 3 (1990), p.33.
6 Max Tegmark, 'What is Reality?', BBC *Horizon*, 17 January 2011.
7 Peter Byrne, 'Everett and Wheeler: The Untold Story', in Saunders et al., p.523.
8 Quoted by Peter Byrne, in Saunders, et al., p.539.
9 Max Tegmark, 'The Interpretation of Quantum Mechanics: Many Worlds or Many Words?', arXiv: quant-ph/9709032 v1, 15 September 1997, p.1.
10 Adrian Kent, 'One World Versus Many: The Inadequacy of Everettian Accounts of Evolution, Probability and Scientific Confirmation', in Saunders, et al., p.309.
11 Greene, *The Hidden Reality*, p.344.
12 Andrei Linde, 'The Self-reproducing Inflationary Universe', *Scientific American*, November 1994, pp.51–2.
13 Raphael Bousso and Leonard Susskind, 'The Multiverse Interpretation of Quantum Mechanics', arXiv: hep-th/1105,3796v1, 19 May 2011, p.2.
14 Alan H. Guth, 'Eternal Inflation and its Implications', arXiv: hep-th/0702178v1, 22 February 2007, pp.9–10.
15 Greene, *The Hidden Reality*, p.91.
16 Ibid., p.155.
17 Susskind, *The Cosmic Landscape*, p.381.

ENDNOTES

18   Justin Khoury, Burt A. Ovrut, Paul J. Steinhardt and Neil Turok, 'The Ekpyrotic Universe: Colliding Branes and the Origin of the Hot Big Bang', arXiv: hep-th/0103239v3, 15 August 2001, pp.3–4. Published in *Physical Review D*, 64, 123522 (2001).
19   Jean-Luc Lehners, Paul J. Steinhardt and Neil Turok, 'The Return of the Phoenix Universe', arXiv: hep-th/0910.0834v1, 5 October 2009, p.4.
20   Robert Adler, 'The Many Faces of the Multiverse', *New Scientist*, 26 November 2011, pp.43 and 47.
21   Greene, *The Hidden Reality*, p.188.

Chapter 10: Source Code of the Cosmos

1   Albert Einstein, 'Motives for Research', speech delivered at Max Planck's sixtieth birthday celebration, April 1918.
2   This was the title of Wigner's Richard Courant lecture in mathematical sciences delivered at New York University on 11 May 1959. It was published in *Communications on Pure and Applied Mathematics*, 13 (1960), pp.1–14.
3   Max Tegmark, 'The Mathematical Universe', *Foundations of Physics*, 38 (2008), p.101: arXiv: gr-qc/0704.0646v2, 8 October 2007, p.1.
4   Interview with Adam Frank: 'Is the Universe Actually Made of Math?', *Discover*, July 2008, published online 16 June 2008: http://discovermagazine.com/2008/jul/16-is-the-universe-actually-made-of-math.
5   See Deutsch, pp.200 and 216.
6   See http://ibmquantumcomputing.tumblr.com/.
7   There are many online versions of Shakespeare's 'Scottish play'. See, for example, http://shakespeare.mit.edu/macbeth/full.html.
8   See Seth Lloyd, 'Computational capacity of the universe', arXiv: quant-ph/0110141v1, 21 October 2001.
9   Hawking, p.105.
10   Quoted by Susskind, *The Black Hole War*, p.185.
11   Susskind, *The Black Hole War*, p.254.
12   Ibid., p.241.
13   Ibid., p.419.
14   The bet, and Hawking's comment, are reproduced in ibid., p.445.
15   Greene, *The Hidden Reality*, p.261.

Chapter 11: Ego Sum Ergo Est

1   Letter to Hendrik Lorentz, 3 February 1915.
2   Fred Hoyle, 'The Universe: Past and Present Reflections', *Annual Review of Astronomy and Astrophysics*, 20 (1982), p.16.
3   Brandon Carter, 'Large Number Coincidences and the Anthropic Principle in Cosmology', in M. S. Longair (ed.), *Confrontation of Cosmological Theories with Data*, Riedel, Dordrecht, 1974, p.127.
4   Leonard Susskind, 'The Anthropic Landscape of String Theory', arXiv: hep-th/0302219v1, 27 February 2003, p.1.

5   Susskind, *The Cosmic Landscape*, p.14.
6   Brandon Carter, 'Large Number Coincidences' , op cit., p.129.
7   http://www.discovery.org/about.php.
8   Helge Kragh, *Centaurus*, 39 (1987), pp.191–4. This quote is reproduced in Kragh, *Higher Speculations*, p.249.
9   Kragh, *Higher Speculations*, p.217.
10 John Templeton, *The Philanthropic Vision of Sir John Templeton*, p.6 (see http://www.templeton.org/sir-john-templeton/philanthropic-vision).
11 Davies, *The Goldilocks Enigma,* p.302.
12 2006 Templeton Prize Chronicle, p.4 (see http://www.templetonprize.org/downloads.html#barrow),
13 'Martin Rees wins 2011 Templeton Prize', press release, 6 April 2011, p.2 (see http://www.templetonprize.org/pdfs/2011_prize/TP-2011-Press-Release.pdf).
14 Ibid., p.3.
15 Interview with Ian Sample, *Guardian,* 6 April 2011, http://www.guardian.co.uk/science/2011/apr/06/astronomer-royal-martin-rees-interview.
16 Bostrom, p.6.
17 Steven Weinberg, 'Anthropic Bound on the Cosmological Constant', *Physical Review Letters*, 59 (1987), p.2607.
18 Helge Kragh, 'An Anthropic Myth: Fred Hoyle's Carbon-12 Resonance Level', *Archive for History of Exact Sciences*, 64 (2010), p.721. I'm grateful to Professor Kragh for drawing this paper to my attention.
19 Lee Smolin, 'Scientific Alternatives to the Anthropic Principle', arXiv: hep-th/0407213v3, 29 July 2004, p.26.
20 Bostrom, p.189.
21 Susskind, *The Cosmic Landscape*, p.357.

Chapter 12: Just Six Questions

1   Albert Einstein, 'Geometry and Experience', Prussian Academy of Sciences, Berlin, 27 January 1921.
2   Alfred J. Ayer (ed.), *Logical Positivism*, The Library of Philosophical Movements, The Free Press of Glencoe, 1959, p.8.
3   David Hume, *An Enquiry Concerning Human Understanding*, Section XII, Part III, http://ebooks.adelaide.edu.au/h/hume/david/h92e/chapter12.html.
4   In Hoddeson, et al., p.508.
5   Felix Salmon, 'Recipe for Disaster: The Formula that Killed Wall Street', *Wired Magazine*, 23 February 2009, http://www.wired.com/techbiz/it/magazine/17-03/wp_quant.
6   Greene, *The Hidden Reality*, p.317.
7   Ibid., p.319.
8   Susskind, *The Cosmic Landscape*, pp.192–3.
9   Lee Smolin and Leonard Susskind, 'Smolin vs Susskind: The Anthropic Principle', *The Edge*, 18 August 2004, http://www.edge.org/3rd_culture/smolin_susskind04/smolin_susskind.html.

ENDNOTES

10 Quoted by Lawrence Krauss, Isaac Asimov Memorial Panel Debate, Hayden Planetarium, American Museum of Natural History, New York, 13 February 2001. Quoted in Woit, p.180.

11 Communication to the author, 6 February 2011.

12 Weinberg, p.133.

13 Smolin and Susskind, 'Smolin vs Susskind', op. cit.

14 Nancy Cartwright and Roman Frigg, 'String Theory Under Scrutiny', *Physics World*, September 2007, p.15.

15 Horgan, p.91.

16 In his excellent 2007 biography of Albert Einstein, Walter Isaacson explained that he could find no direct evidence that Kelvin had made this pronouncement. Nevertheless, the statement captures something of the mood that prevailed among prominent physicists at this time.

17 Attributed to Lord Kelvin. Quoted in Isaacson, p.90, but see also the footnote on p.575.

18 Peter Woit, Not Even Wrong, blog entry, 27 June 2011. http://www.math. columbia.edu/~woit/wordpress/?p=3811.

Bibliography

Aczel, Amir D., *Entanglement: The Greatest Mystery in Physics*, John Wiley & Sons Ltd., Chichester, 2003.

Baggott, Jim, *A Beginner's Guide to Reality*, Penguin, London, 2005.

Baggott, Jim, *The Quantum Story: A History in 40 Moments*, Oxford University Press, 2011.

Baggott, Jim, *Higgs: The Invention and Discovery of the 'God Particle'*, Oxford University Press, 2012.

Barbour, Julian, *The End of Time,* Weidenfeld & Nicholson, London, 1999.

Barrow, John D., *Theories of Everything,* Vintage, London, 1991.

Barrow, John D., *The Constants of Nature: From Alpha to Omega*, Vintage, London, 2003.

Barrow, John D., *The Book of Universes*, Vintage, London, 2012.

Barrow, John D., and Tipler, Frank, *The Anthropic Cosmological Principle*, Oxford University Press, 1986.

Bell, J. S., *Speakable and Unspeakable in Quantum Mechanics,* Cambridge University Press, 1987.

Beller, Mara, *Quantum Dialogue,* University of Chicago Press, 1999.

Bernstein, Jeremy, *Quantum Profiles,* Princeton University Press, 1991.

Bostrom, Nick, *Anthropic Bias: Observation Selection Effects in Science and Philosophy*, Routledge, New York, 2002.

Brooks, Michael, *Free Radicals: The Secret Anarchy of Science*, Profile Books, London, 2011.

Calaprice, Alice, *The Ultimate Quotable Einstein*, Princeton University Press, 2011.

Carroll, Sean, *From Eternity to Here: The Quest for the Ultimate Theory of Time*, Oneworld Publications, Oxford, 2011.

Cassidy, David C., *Uncertainty: The Life and Science of Werner Heisenberg.* W. H. Freeman, New York, 1992.

Chalmers, A. F., *What is This Thing Called Science?* (3rd edition), Hackett, Indianapolis, 1999.

BIBLIOGRAPHY

Close, Frank, *The Infinity Puzzle*, Oxford University Press, 2011.

Cox, Brian, and Forshaw, Jeff, *Why Does E = mc²?* Da Capo Press, Cambridge, MA, 2009.

Crease, Robert P., and Mann, Charles C., *The Second Creation: Makers of the Revolution in Twentieth-century Physics*, Rutgers University Press, 1986.

Crease, Robert P., *A Brief Guide to the Great Equations: The Hunt for Cosmic Beauty in Numbers*, Robinson, London, 2009.

Cushing, James T., *Philosophical Concepts in Physics*, Cambridge University Press, 1998.

Davies, P. C. W., and Brown, J. R. (eds.), *The Ghost in the Atom,* Cambridge University Press, 1986.

Davies, Paul, *God and the New Physics*, Penguin Books, 1984.

Davies, Paul, *Other Worlds: Space, Superspace and the Quantum Universe*, Penguin Books, 1988.

Davies, Paul, *About Time: Einstein's Unfinished Revolution*, Penguin Books, 1995.

Davies, Paul, *The Goldilocks Enigma: Why is the Universe Just Right for Life?,* Allen Lane, London, 2006.

Davies, Paul, and Gregersen, Niels Henrik, *Information and the Nature of Reality: From Physics to Metaphysics*, Cambridge University Press, 2010.

Deutsch, David, *The Fabric of Reality,* Penguin, London, 1997.

DeWitt, Bryce. S., and Graham, Neill (eds.), *The Many Worlds Interpretation of Quantum Mechanics,* Pergamon, Oxford, 1975.

Dyson, Freeman, *Disturbing the Universe*, Basic Books, New York, 1979.

Farmelo, Graham (ed.), *It Must be Beautiful: Great Equations of Modern Science*, Granta Books, London, 2002.

Farmelo, Graham, *The Strangest Man: The Hidden Life of Paul Dirac, Quantum Genius*, Faber and Faber, London, 2009.

Feyerabend, Paul, *Against Method* (3rd edition), Verso, London, 1993.

Feynman, Richard P., *QED: The Strange Theory of Light and Matter*, Penguin, London, 1985.

Fine, Arthur, *The Shaky Game: Einstein, Realism and the Quantum Theory* (2nd edition), University of Chicago Press, 1986.

French, A. P., *Special Relativity*, Van Nostrand Reinhold, Wokingham, 1968.

French, A. P., and Kennedy, P. J. (eds.), *Niels Bohr: a Centenary Volume,* Harvard University Press, Cambridge, MA, 1985.

Gamow, George, *Thirty Years that Shook Physics,* Dover Publications, New York, 1966.

Gell-Mann, Murray, *The Quark and the Jaguar,* Little, Brown & Co., London, 1994.

Gillies, Donald, *Philosophy of Science in the Twentieth Century*, Blackwell, Oxford, 1993.

Gleick, James, *Genius: Richard Feynman and Modern Physics,* Little, Brown & Co., London, 1992.

BIBLIOGRAPHY

Gleick, James, *Isaac Newton*, Harper Perennial, 2004.

Greene, Brian, *The Elegant Universe: Superstrings, Hidden Dimensions and the Quest for the Ultimate Theory*, Vintage Books, London, 2000.

Greene, Brian, *The Fabric of the Cosmos: Space, Time and the Texture of Reality*, Allen Lane, London, 2004.

Greene, Brian, *The Hidden Reality: Parallel Universes and the Deep Laws of the Cosmos*, Allen Lane, London, 2011.

Gribbin, John, *Q is for Quantum: Particle Physics from A to Z*, Weidenfeld & Nicolson, London, 1998.

Gribbin, John, *In Search of the Multiverse*, Allen Lane, London, 2009.

Gubser, Stephen S., *The Little Book of String Theory*, Princeton University Press, 2010.

Guth, Alan H., *The Inflationary Universe: The Quest for a New Theory of Cosmic Origins*, Vintage, London, 1998.

Hacking, Ian, *Representing and Intervening*, Cambridge University Press, 1983.

Halpern, Paul, *Collider: The Search for the World's Smallest Particles*, John Wiley & Son, Inc., New Jersey, 2009.

Hawking, Stephen, *A Brief History of Time: From the Big Bang to Black Holes*, Transworld Publishers, London, 1988.

Hawking, Stephen, and Mlodinow, Leonard, *The Grand Design: New Answers to the Ultimate Questions of Life*, Bantam Press, London, 2010.

Heisenberg, Werner, *Physics and Philosophy: The Revolution in Modern Science,* Penguin, London, 1989 (first published 1958).

Hoddeson, Lillian, Brown, Laurie, Riordan, Michael, and Dresden, Max, *The Rise of the Standard Model: Particle Physics in the 1960s and 1970s*, Cambridge University Press, 1997.

Horgan, John, *The End of Science: Facing the Limits of Knowledge in the Twilight of the Scientific Age*, Little, Brown & Co., 1997.

Isaacson, Walter, *Einstein: His Life and Universe*, Simon & Shuster, New York, 2007.

Johnson, George, *Strange Beauty: Murray Gell-Mann and the Revolution in Twentieth-Century Physics*, Vintage, London, 2001.

Kane, Gordon, *Supersymmetry: Unveiling the Ultimate Laws of the Universe*, Perseus Books, Cambridge, MA, 2000.

Kennedy, J. B., *Space, Time and Einstein: An Introduction*, Acumen, Chesham, 2003.

Koestler, Arthur, *The Sleepwalkers: A History of Man's Changing Vision of the Universe*, Penguin, London, 1964.

Kragh, Helge, *Quantum Generations: A History of Physics in the Twentieth Century*, Princeton University Press, 1999.

Kragh, Helge, *Higher Speculations: Grand Theories and Failed Revolutions in Physics and Cosmology*, Oxford University Press, 2011.

Kuhn, Thomas S., *The Structure of Scientific Revolutions* (2nd edition), University of Chicago Press, 1970.

BIBLIOGRAPHY

Kumar, Manjit, *Quantum: Einstein, Bohr and the Great Debate About the Nature of Reality*, Icon Books, London, 2008.

Lederman, Leon, with Teresi, Dick, *The God Particle: If the Universe is the Answer, What is the Question?*, Bantam Press, London, 1993.

Mehra, Jagdish, *The Beat of a Different Drum: The Life and Science of Richard Feynman*, Oxford University Press, 1994.

Mehra, Jagdish, *Einstein, Physics and Reality*, World Scientific, London, 1999.

Moore, Walter, *Schrödinger: Life and Thought*, Cambridge University Press, 1989.

Nambu, Yoichiro, *Quarks*, World Scientific Publishing, Singapore, 1981.

Nola, Robert, and Sankey, Howard, *Theories of Scientific Method*, Acumen, Durham, 2007.

Overbye, Dennis, *Lonely Hearts of the Cosmos: The Quest for the Secret of the Universe*, Picador, London, 1993.

Pais, Abraham, *Subtle is the Lord: The Science and the Life of Albert Einstein*, Oxford University Press, 1982.

Pais, Abraham, *Niels Bohr's Times, in Physics, Philosophy and Polity*, Clarendon Press, Oxford, 1991.

Panek, Richard, *The 4% Universe: Dark Matter, Dark Energy and the Race to Discover the Rest of Reality*, Oneworld, Oxford, 2011.

Penrose, Roger, *Cycles of Time: An Extraordinary New View of the Universe*, Vintage, London, 2011.

Popper, Karl R., *The Logic of Scientific Discovery*, Hutchinson, London, 1959.

Popper, Karl R., *Quantum Theory and the Schism in Physics*, Unwin Hyman, London, 1982.

Psillos, Stathis, *Scientific Realism: How Science Tracks Truth*, Routledge, London, 1999.

Psillos, Stathis, and Curd, Martin (eds.), *The Routledge Companion to Philosophy of Science*, Routledge, London, 2010.

Rae, Alastair, *Quantum Physics: Illusion or Reality?*, Cambridge University Press, 1986.

Randall, Lisa, *Warped Passages: Unravelling the Universe's Hidden Dimensions*, Penguin Books, London, 2006.

Randall, Lisa, *Knocking on Heaven's Door: How Physics and Scientific Thinking Illuminate the Universe and the Modern World*, Random House, London, 2011.

Ray, Christopher, *Time, Space and Philosophy*, Routledge, London, 1991.

Rees, Martin, *Just Six Numbers: The Deep Forces that Shape the Universe*, Phoenix, London, 2000.

Reichenbach, Hans, *The Direction of Time*, Dover, New York, 1999 (first published 1956).

Riordan, Michael, *The Hunting of the Quark: A True Story of Modern Physics*, Simon & Schuster, New York, 1987.

Rindler, Wolfgang, *Introduction to Special Relativity*, Oxford University Press, 1982.

Rovelli, Carlo, *Quantum Gravity*, Cambridge University Press, 2007.

BIBLIOGRAPHY

Russell, Bertrand, *The Problems of Philosophy*, Oxford University Press, 1967.

Sample, Ian, *Massive: The Hunt for the God Particle*, Virgin Books, London, 2010.

Saunders, Simon, Barrett, Jonathan, Kent, Adrian, and Wallace, David (eds.), *Many Worlds? Everett, Quantum Theory and Reality*, Oxford University Press, 2010.

Schilpp, Paul Arthur (ed.), *Albert Einstein. Philosopher-scientist*, The Library of Living Philosophers, Volume 1, Harper & Row, New York, 1959 (first published 1949).

Schweber, Silvan S., *QED and the Men Who Made It: Dyson, Feynman, Schwinger, Tomonaga*, Princeton University Press, 1994.

Singh, Simon, *Big Bang: The Most Important Scientific Discovery of All Time and Why You Need to Know About It*, Harper Perennial, London, 2005.

Smolin, Lee, *Three Roads to Quantum Gravity*, Weidenfeld & Nicolson, London, 2000.

Smolin, Lee, *The Trouble with Physics: The Rise of String Theory, the Fall of a Science and What Comes Next*, Penguin, London, 2006.

Stachel, John (ed.), *Einstein's Miraculous Year: Five Papers that Changed the Face of Physics*, Princeton University Press, 2005.

Stannard, Russell, *The End of Discovery*, Oxford University Press, 2010.

Susskind, Leonard, *The Cosmic Landscape: String Theory and the Illusion of Intelligent Design*, Little, Brown and Co., New York, 2006.

Susskind, Leonard, *The Black Hole War: My Battle With Stephen Hawking to Make the World Safe for Quantum Mechanics*, Little, Brown & Co., New York, 2008.

't Hooft, Gerard, *In Search of the Ultimate Building Blocks*, Cambridge University Press, 1997.

Thorne, Kip S., *Black Holes and Time Warps: Einstein's Outrageous Legacy*, W. W. Norton & Company, New York, 1994.

van der Waerden, B. L., *Sources of Quantum Mechanics*, Dover, New York, 1968.

Vedral, Vlatko, *Decoding Reality: The Universe as Quantum Information*, Oxford University Press, 2010.

Veltman, Martinus, *Facts and Mysteries in Elementary Particle Physics*, World Scientific, London, 2003.

Vignale, Giovanni, *The Beautiful Invisible: Creativity, Imagination, and Theoretical Physics*, Oxford University Press, 2011.

Weinberg, Steven, *Dreams of a Final Theory: The Search for the Fundamental Laws of Nature*, Vintage, London, 1993.

Wheeler, John Archibald, and Zurek, Wojciech Hubert (eds.), *Quantum Theory and Measurement*, Princeton University Press, 1983.

Wheeler, John Archibald, with Ford, Kenneth, *Geons, Black Holes and Quantum Foam: A Life in Physics*, W. W. Norton & Company, New York, 1998.

Whitrow, G. J., *What Is Time?*, Oxford University Press, 2003 (first published 1972).

Wilczek, Frank, *The Lightness of Being: Big Questions, Real Answers*, Allen Lane, London, 2009.

BIBLIOGRAPHY

Woit, Peter, *Not Even Wrong: The Failure of String Theory and the Continuing Challenge to Unify the Laws of Physics*, Vintage, London, 2007.

Yau, Shing-Tung, and Nadis, Steve, *The Shape of Inner Space: String Theory and the Geometry of the Universe's Hidden Dimensions*, Basic Books, New York, 2010.

Zee, A., *Fearful Symmetry: The Search for Beauty in Modern Physics*, Princeton University Press, 1999.

Zee, A., *Quantum Field Theory in a Nutshell*, Princeton University Press, 2003.

Index

INDEX